C000241971

# Translocal Ageing in the Global East

Deljana Iossifova

# Translocal Ageing in the Global East

## Bulgaria's Abandoned Elderly

Deljana Iossifova
Urban Studies
University of Manchester
Manchester, UK

ISBN 978-3-030-60822-4     ISBN 978-3-030-60823-1   (eBook)
https://doi.org/10.1007/978-3-030-60823-1

© The Editor(s) (if applicable) and The Author(s) 2020
This work is subject to copyright. All rights are solely and exclusively licensed by the Publisher, whether the whole or part of the material is concerned, specifically the rights of translation, reprinting, reuse of illustrations, recitation, broadcasting, reproduction on microfilms or in any other physical way, and transmission or information storage and retrieval, electronic adaptation, computer software, or by similar or dissimilar methodology now known or hereafter developed.
The use of general descriptive names, registered names, trademarks, service marks, etc. in this publication does not imply, even in the absence of a specific statement, that such names are exempt from the relevant protective laws and regulations and therefore free for general use.
The publisher, the authors and the editors are safe to assume that the advice and information in this book are believed to be true and accurate at the date of publication. Neither the publisher nor the authors or the editors give a warranty, expressed or implied, with respect to the material contained herein or for any errors or omissions that may have been made. The publisher remains neutral with regard to jurisdictional claims in published maps and institutional affiliations.

Cover pattern © Harvey Loake

This Palgrave Macmillan imprint is published by the registered company Springer Nature Switzerland AG.
The registered company address is: Gewerbestrasse 11, 6330 Cham, Switzerland

# Acknowledgements

Many people have helped to develop this book over the past five or so years. First and foremost, my thanks are due to the older people in Bulgaria who agreed to share their time, life histories and everyday lives as participants in the research.

Many thanks to the University of Manchester for the small grant that allowed me to do the project in the first place; the extraordinarily talented Elena Balabanska, who worked with me in the field; Alan Lewis, who provided helpful insights at the start of the project; Evgeny Bely and the devoted social workers who allowed me to be part of their communities for a short while; and the generous Ginka Hadjieva, who introduced me to many of the older people whose lives are the subject of this book. Thanks to colleagues at the Antwerp Urban Studies Institute and Stijn Oosterlynck, in particular, for stimulating conversations during the early stages of research.

My special thanks are due to the inspirational Clara Greed, Anna Plyushteva and Tim Schwanen who read the full manuscript and offered very helpful critical feedback and advice. The exceptionally brilliant Qi Liu and Yahya Gamal provided comments that helped me see the forest for the trees. Thanks also to the remarkable Sarah Marie Hall, as well as Debapriya Chakrabarti and Joe Shaw, who all commented on drafts of the chapters. Thanks to Ali Browne—for being fabulous; introducing me, probably unknowingly, to the most interesting of literatures; and, of course, for letting us stay in her house during the pandemic so that we had room to breathe, work and play.

The book was commissioned by Palgrave's Holly Tyler and seen through to publication by Joshua Pitt—I am grateful to both for their patience and unbroken support and understanding. Thanks also to Sophie Li in Shanghai.

Thank you to my family. Thank you to Alma for keeping me company throughout this journey, even before she was born; and to Amaya for smiling and laughing, as she so often does, to brighten up dark Manchester days.

Most of all, thank you to my mother. Without her, this book would have never materialised. She was part of the project in more ways than I can list here, introducing me to friends and acquaintances during the initial stages of this research; offering insights into her life and experiences under state socialism and after; and serving as a critical friend as I worked through my interpretations of the material. So that I had space to think and write, she helped looking after her granddaughters in the most fervent and generous of ways, teaching them, in the process, how to be fierce and fearless just like her. This book is for her.

# CONTENTS

# List of Figures

# CHAPTER 1

# Introduction: Ageing in Bulgaria

**Abstract** In this chapter, Iossifova offers a brief introduction to Bulgaria's recent history and the political and socioeconomic transitions that have led to the main challenges facing the country today: a rapidly shrinking and ageing population due to the unprecedented outmigration of younger generations in search of better livelihoods. Iossifova reviews the recent literature on ageing in human geography and related disciplines, including notions of 'ageing in place', age-relationality and the frameworks of intersectionality, intergenerationality and the lifecourse. She introduces the human ecosystem framework and discusses the grounded theory approach taken, including interviews, observation and autoethnography. She then moves on to present the case study areas in Sofia and the Village in the Bulgarian Balkans. She closes in outlining the structure of the book.

**Keywords** Bulgaria • Ageing population • Human ecosystem framework • 'Ageing in place' • Age-relationality • Global East • Post-socialism

I grew up in Sofia. I was the youngest member of a household which, at that time, consisted of my father, mother, older brother and maternal grandmother. We lived in a one-bedroom apartment: my brother and I slept in the bedroom, my parents on a sofa bed in the living room and my grandmother in the kitchen. My parents would leave for work and my

© The Author(s) 2020
D. Iossifova, *Translocal Ageing in the Global East*,
https://doi.org/10.1007/978-3-030-60823-1_1

brother for school early in the mornings. Very occasionally, I would go to kindergarten, but that was the exception. I knew that if I cried loudly enough once my father began getting me ready, at least one of our elderly neighbours would come rushing through the door and insist to look after me for the rest of the day. My playmates were kids my age who, like me, were looked after by their grandparents. Grandmothers were in charge of cooking lunches and dinners and chasing us with snacks in-between. Apartment doors were never locked and we could walk in and out of neighbours' homes as we pleased.

At the time, my grandmother was in her early 80s and quite sick. She had trouble walking, but she didn't need to walk much because everything she may have needed was taken care of by my parents and neighbours. She would spend her days watching TV in the living room, sharing the news with neighbours on the bench outside our building or sitting in the kitchen chatting away with frequent visitors. She would get her pension delivered once a month, a welcome occasion for a cup of coffee with the officer delivering it; as was receiving a letter from the mail man or woman; the reading of the water or electricity meters; or the refuelling of petroleum stocks for the winter months. Naturally, all of my mother's four siblings would visit at least a couple of times a month, and so would their adult children. Our apartment was never quiet.

Neither was the building or surrounding areas. Bounded by tall birch trees, there was a huge playground on one side of it, and small green areas on the other with linden trees that seemed to be losing their pea-like fruit year-round. There was also a kind of parking lot covered by the slender branches of remarkably tall willow trees. Children and teenagers found miraculous ways of occupying all of these spaces during the day, likely under their grandparents' unnoticed watchful gaze. At times, the entire population of all buildings in the street would congregate outside to roast peppers in view of canning for the winter months; to beat the rugs, a practice now forgotten; or to pass time when electricity was down again and there was nothing else to do but to 'click' roasted pumpkin or sunflower seeds and chat.

*    *    *

My paternal grandmother, an opera singer, divided her time between Sofia and the Village, officially a town but actually a village hidden away in the Bulgarian Balkans. She did her morning exercises upon waking up to keep in shape and had a strict beauty routine which included the application of face cream, mornings and evenings, on every single day. She insisted on dying her hair until (almost) the day she died. I spent my summers with her in the Village. She taught us grandchildren how to sit straight 'like a lady', how to make rice pudding and how to work the soil. Having lived through the war, she believed in the importance of knowing how to grow food and tirelessly maintained her father's house and the rather huge garden attached to it in the Village.

Together with my grand-aunt, she hosted all four grandchildren during the summer months (and, in the case of my pubescent brother, during his wildest teenage years). Of similar age, we were probably quite easy to look after—bread and butter in the mornings, a light lunch and cheese and olives in the evenings. In-between, we ran off to explore the forests, rummage through the cellar or attic of the old house, water the plants in the garden or feed the neighbour's pigs. In the cool evenings, we put on sweaters and followed my grandmother up the street and then the hundreds of steps to the monastery, stopping here and there to wait for others to join us or for my grandmother to exchange the latest gossip. Once at the top, at the foot of the glorious church, we would chase the ribbiting frogs or sit quietly and listen to the older people's conversations.

On especially hot days, my grandmother kept us indoors. Then, I lay on the divan in the kitchen (which, of course, also served as the bedroom for my grandmother or grand-aunt when their house was invaded by the younger generations), read one of the many books I had retrieved from the cellar or attic, and watched my grandmother rumble about, canning vegetables or making jam.

*　*　*

This book is about ageing in Bulgaria. Bulgaria's recent political and socioeconomic transitions have led citizens to reconsider fundamental questions, such as whether or not to have children or whether to stay in their home country at all. The processes of pervasive emigration and shrinking fertility rates have become entangled with the rapid ageing of the population. Nearly 20% of Bulgarians are now 65 years and over. The old-age dependency ratio (the ratio of those 65 years and older to the labour force, i.e., those 15 to 64) is estimated at 30.4 and the country's median age is now 42.6 years (Central Intelligence Agency 2017). Bulgaria's transition to market economy has accommodated and encouraged changes in the role of older people in the family. The literature observes the fading of traditionally strong family ties in Balkan countries (considered closer to Asian, rather than Western, family types [Goody 1996]), dwindling intergenerational dialogue and increased focus on the individual (Kulcsár and Brădățan 2014). Today, older people are often associated with the communist past and rendered responsible for the difficulties of the present, suggesting a generational divide and a series of challenges to the provision of services for the elderly.

With this book I seek to draw attention to the rapid demographic, socioeconomic, cultural and ecological transitions taking place in a corner on the very periphery of Europe. We know a lot about the person-environment relationship in urban and rural environments, yet the literature focuses predominantly on the experiences of older people in the Global North (Smith et al. 2004; Scharf 2003; Scharf et al. 2005; Buffel et al. 2012, 2013). Work on ageing in the Global East—comprising those societies that are 'too rich to be in the South, too poor to be in the North' (Müller 2020)—is only beginning to emerge.

Of course, the paragraphs at the start of this chapter are tinted with nostalgia—the feeling of having lost something valuable, 'a positively toned evocation of a lived past in the context of some negative feeling toward present or impending circumstance' (Davis 1979; see also Ghodsee 2004). Capitalism and globalisation have worked their parts to create expectations of and cookie-cut patterns for the lives we want to live today, regardless of where in the world we are located. Such expectations do not seem to allow the required space nor time for the type of intergenerational relations that I grew up with.

Sharing a one-bedroom apartment across three generations, as my family did, may well not be everyone's cup of tea; however, thinking back, it strikes me as exceptional that my parents never had to worry about everyday childcare (which was free) or how to look after us during school holidays (there would always be some relative to send the children off to in the countryside). Bulgaria's urban population maintained close ties with the countryside—*provintsiyata* (the 'province'), a term with pejorative connotations commonly used to refer to areas outside the country's capital (Staddon 2004).

Neither would it have crossed my parents' minds to put my ageing grandparents in a nursing home. Even when we eventually moved country and left them behind, we could be sure they would be well taken care of by other members of the family. This is associated with the formerly predominant cultural model of family relations and the provision of care for older people at home and by family members. However, signalling a break with the traditional model of care for the elderly, demand for nursing home places has been steadily growing over the past decades, with 7 men and 18 women on the waiting list for each of the few available places (Gancheva and Chengelova 2006). Bulgaria's population-to-nursing homes ratio continues to be abysmally low (there were only 43.85 nursing and elderly home beds per 100,000 people in 2014, compared with the European average of 762.69) (WHO 2019).

As global academic communities explore notions of healthy cities, age-friendly cities, 'ageing in place' and a full range of related ideas situated in more affluent settings, research is focused to a great extent on age-friendly features in the built environment. As an architect with years of experience in practice I have no doubt that the quality of the built environment plays a decisive role when it comes to levels of human wellbeing, especially for the elderly and vulnerable. However, the ageing population in Bulgaria, one of the world's top five most rapidly ageing countries, is facing challenges of a different calibre.

In this book, I ask how Bulgaria's dramatic socio-political transitions over the past eight or so decades have influenced the lifecourse of older people in the country today. I carefully unravel their patterns of everyday life in the capital Sofia as well as a village ('the Village') in the Bulgarian Balkans in order to draw out the mechanisms through which they manage meagre pensions, sustain their ailing bodies and make do in tattered homes. I ask how they adapt to their continuously shifting environment and a state of permanent uncertainty. Finally, I argue that 'ageing in place', a popular notion and policy agenda, does not fit the realities or expectations of ageing in Bulgaria or the wider context of the Global East. Beyond drawing attention to a marginalised research context, I suggest that in times of socioeconomic, environmental, health and climate crises, it is time to adopt methodological and theoretical approaches that transcend traditional disciplinary or domain divides in order to uncover, capture and understand the complexity of ageing, entangled with other processes, in rapidly changing environments.

My contribution is not the explicit comparison between ageing in the Global East and ageing in other socio-geographical contexts. Indeed, within the space of this short book, I sometimes merely touch upon aspects that deserve in-depth attention. Therefore I hope that the realities presented here provide inspiration for future research and creative interventions.

## GLOBAL TRANSFORMATIONS, BULGARIAN CHALLENGES

Between 1944 and 1989, Bulgaria was under Communist rule. On 9 September 1944 (a significant date representing the transition and commonly referred to as *deveti*—'the ninth'), the Bulgarian Communist Party (BCP) de facto ended Bulgarian monarchy in a coup supported by the Soviets. The longest-serving Head of State in the Soviet Bloc, Todor Zhivkov, presided over a comparatively benign totalitarian regime between 1956 and 1989. The newly established People's Republic introduced a centrally planned economy, transforming the previously agrarian into a moderately successful industrialised state with little tolerance of alternative parties, religious organisations and, towards the end of the Communist era, some ethnic minorities (Crampton 2005).

The regime began to crumble in the late 1980s and eventually broke down in November 1989, when Zhivkov was removed following similar events in other Eastern Bloc countries. The decades following the fall of the socialist system were marked by economic restructuring and political instability (exemplified by five different Prime Ministers and even more governments in the first half of the 1990s). The shut-down of state-owned enterprises in the wake of political change in the late 1980s and early 1990s led to an unprecedented rise in unemployment rates which today are still higher in Bulgaria than the European average (World Bank 2017). As is often the case, changes in the political structure of the country led to transitions in the lives of individuals who are now faced with previously unknown freedoms as well as risks: unemployment, poverty, social exclusion, limited access to health care (Hoff and Perek-Bialas 2015).

Bulgaria joined the European Union (EU) in 2007. In 2019, 22.6% of its total population were at risk of poverty (and 34.6% of people over 65) (eurostat 2020). During the same year, its ever-shrinking population fell below seven million for the first time since 1945 (NSI 2020). Almost 95% of all cities in Bulgaria are losing their population (Central Intelligence Agency 2017). Sofia (the capital), Plovdiv, Varna and a handful of cities along the Black Sea coast corridor are rare exceptions. Shrinking cities continuously lose social and economic functions and services, leading to the decline of the urban environment and subsequent outmigration (Pallagst et al. 2009). This shrinking is the result of a combination of radical shifts in the political system, deindustrialisation, suburbanisation and population ageing (Oswalt 2005).

Yet, whilst Sofia draws people from other cities, in turn these cities draw people from the countryside (Staddon 2004). This is because young adults in rural areas face fundamental difficulties in finding employment and alternatives to rural-to-urban migration hardly exist (Tsekov 2017). Between 1990 and 2015, Bulgaria's countryside lost 40% of its population (World Bank 2017). Its villages are deserted. For instance, a village some 150 km from Sofia saw its population drop from over 500 to 17 in only a few years. The average age there is well over 75. An 80-year-old woman reports that her family had left, 'like everyone else', to find jobs. The closest village now counts a population of three; another village close-by has turned into a ghost town (Evans 2013).

A considerable number of senior citizens are left behind as younger generations seek to secure livelihoods elsewhere, including abroad (King et al. 2017). Migration and ageing are already two key processes reshaping the structure and distribution of Europe's population (Chłoń-Domińczak et al. 2014; King and Lulle 2016). Yet research examining the experiences of those left behind in the migration process is rather scarce. Here, too, studies focus on higher-income countries or are concerned with rural settings in lower-income countries. Whether or not transnational care is available has impact on the vulnerability of parents left behind (Bastia et al. 2020). Whilst we know about the financial support provided to parents, we know significantly less about the challenges they face and the coping mechanisms they develop (Conkova et al. 2019).

Bulgaria's challenges are not exceptional. The country shares much in common with other countries of the Global East (Müller 2020)—a notion newly capturing the scholarly imagination to complement emerging recognition of the need to decolonise scholarship and rethink established categories. Müller (2020) delimits the reach of the Global East to 'those societies that took part in what was the most momentous global experiment of the twentieth century: to create communism'. Characterising the Global East is its 'Eastness', the being in the 'interstices between North and South', 'outside the circuits and conduits of Western knowledge architecture' and 'in the middle of global relations rather than cut off from them' (Müller 2020). Exploring the 'multiple experiences—of empire, globalisation, neoliberal reform, nationalist populism, political resistance, asymmetric wars' (Müller 2020) of the East allows us to connect it with the Global elsewhere.

Global demographic trends show that the future is urban (as is overly well known) and 'grey'. Increasing proportions of elderly people will require care and services. Ageing urban populations are likely going to be living with vastly different and rapidly changing economic, political and—crucially—environmental conditions. Climate change is already triggering processes that expose previously unthinkable vulnerabilities in the context of the Global North, for instance, in the built environment (e.g., heat waves or increased risk of flooding) (The Royal Society 2012). Economic and longer-term migration impacts on processes of identity formation and maintenance, family structure, adaptation and integration. Extreme weather conditions are affecting human health and well-being, and resource scarcities are linked with the need to alter almost every aspect of human life: how we move and travel, what and how we eat, how we communicate, and how we sanitise our bodies and homes (made painfully explicit by the currently raging COVID-19 pandemic). In this sense, thinking about the future requires a good measure of thinking about a great range of interrelated systems that interact in multiple and frequently unpredictable ways.

## From 'Ageing in Place' to Age-Relationality

In the following section, I provide an overview of approaches in human geography and related disciplines that conceptualise ageing in space and place. I describe ideas of 'ageing in place', active ageing and the associated 'age-friendly cities' initiative (WHO 2007), and draw out why a relational approach (building on an age-relationality framework) is central to research on ageing.

In the literature, ageing is defined as the process of irreversible change in the structure of an organism, accompanied by the gradual diminishing of physical and psychological functions and disturbed adaptation to the living environment (Zhecheva 2007). Whilst inevitable, ageing is a process that is highly dependent on a whole range of factors, internal and external, that make it impossible to create a universal theory of ageing. Internal factors include, for instance, the overall health of a person or their mental health status. They include their access to financial, material and social resources and networks and their ability (and willingness) to participate in activities. External factors include the quality of the living environment or the provision and governance of different kinds of infrastructures (such as health care or transportation). How people experience the ageing process is therefore linked with transitions in their identities, abilities and needs which, in turn, are embedded in a continuously changing environment (see Hörschelmann 2011).

Geographies of ageing encompass work undertaken in health geography, population geography and social geography as well as within the relatively recent interdisciplinary field of geographical gerontology (Andrews et al. 2007, 2009). At the core of such geography-rooted scholarship is the people-environment relationship, studied broadly along the following three axes: (a) the spatial distribution and movement of the elderly; (b) health, care and caregiving and (c) the settings and environments within which ageing occurs (Skinner et al. 2015). Here, research is concerned with the ways in which older people are supported (or challenged) in specific spatial environments, such as residential care environments, neighbourhoods, cities or villages (e.g., Smith 2009b; Phillips et al. 2004; Manthorpe et al. 2004).

The conflation of two rapid processes, global urbanisation and population ageing, has led to the urgent need to rethink how urban communities are positioned to support older citizens. Generally, urban areas are thought to be advantageous in this respect: they offer more immediate access to amenities and services; larger likelihoods to access specialist resources for minority groups (Buffel and Phillipson 2011); access to a wider range of social networks, ranging from neighbours to family and friends (Gardner 2011); and first-hand opportunities to engage with innovation, such as developments in the field of digital technologies (Ratti and Townsend 2011).

However, the combined effect of globalisation, urban regeneration and austerity likely impacts significantly on the experience of older people. Locally, this means that social programmes are impacted by financial cuts, that the characteristics of urban development (gentrification, for instance, or urban sprawl) put pressure on older people in the city and that the privatisation of urban space may well have repercussions for the elderly. As the physical and social context of cities is transformed by such forces, older people are often left to their own devices (Buffel and Phillipson 2016). 'Daily hassles' (Phillips et al. 2004), such as overcrowding or air pollution, become obstacles to ageing well. Cities become difficult to navigate where the environment is difficult to manage (e.g., hilly, uneven) or facilities, such as public toilets, cannot be easily found (e.g., Greed 2019).

Therefore, policy and planning agendas have adopted an ageing lens in order to adapt cities to shifts in contemporary society (Golant 2014). For instance, the concept of age-friendly cities was developed by the World

Health Organization (WHO) with the aim to support and encourage 'active ageing' as 'the process of optimizing opportunities for health, participation and security in order to enhance quality of life as people age' (WHO 2007). The ideal age-friendly city supports active ageing through a variety of mechanisms, such as recognising older people's capacities and resources, anticipating their ageing-related needs, respecting differing lifestyle choices, protecting the vulnerable and promoting the inclusion of older people and their contribution to community life. To make the concept applicable in practice, the WHO developed a checklist of characteristics against which a city's age-friendliness can be measured. These include certain standards for outdoor spaces and buildings, transportation, housing, social participation, respect and social inclusion, civic participation and employment, communication and information, and age-friendly community and health services. In this sense, active ageing is understood to depend much on material conditions and social factors and their interaction (WHO 2007).

The nature of the relationship between older people and the spaces and places within which they age is understood to be mutually constitutive and transactional in nature (Cutchin 2009). However, the ideology of 'ageing in place' fails to address older people's heterogeneity across the axes of culture, gender, socioeconomic status, health and geographic location. The 'ageing in place' narrative perpetuates the normative assumption of a desire for continuity or independence among all ageing people (Estes 2004). Allison E. Smith (2009) proposes a model of the person-environment relationship that incorporates physical attachment and area knowledge, social attachment, historical attachment, religiosity and spirituality, the life history and public spaces. This work suggests that ageing should be studied using a more comprehensive framework which helps to incorporate (and relate to one another) multiple factors—or their proxies.

As famously posed in actor-network-theory, social phenomena, natural phenomena and any discourse about them should be considered a part of a hybrid whole fabricated by the interaction of actors, things and our conceptualisation of the world (Latour 2005). Recent work in human geography calls for a shift towards non-representational and relational thinking and research which accounts for the 'full range of the experiences of ageing' (Skinner et al. 2015) and enables a better understanding of the dynamic people-environment relationship. Non-representational thinking is concerned with the 'description of the bare bones of actual occasions' (Thrift 2007). Similar to principles in actor-network-theory (Latour

2005), non-representational theories assign equal importance to humans and the non-human and are concerned with the ways in which 'life takes shape and gains expression in shared experiences, everyday routines, fleeting encounters, embodied movements' and so on (Lorimer 2005).

Approaches in relational geography understand that 'all things found in or associated with particular spaces and places ... also have connections to things in other spaces and places, often at different scales' (Skinner et al. 2015); furthermore, spaces and places are never considered complete, but always becoming (Hörschelmann 2011; also Iossifova 2015). Relational research on the particular infrastructures that support older people— material and in the shape of services—is emerging (Cloutier-Fisher and Joseph 2000; Peace et al. 2005; Hodge 2008). For instance, a recent body of work along these lines examines the interactions between mobility, wellbeing and independence in ageing (Schwanen and Páez 2010; Schwanen and Ziegler 2011; Ziegler and Schwanen 2011; Schwanen et al. 2012a, b; Nordbakke and Schwanen 2014; Plyushteva and Schwanen 2018). However, there is urgent need to develop our understanding in this field further in order to inform policies and interventions that support older adults in the future (Skinner et al. 2015).

Although relational thinking is relatively underrepresented in research on ageing, debates around 'age-relationality' are beginning to form. Discussing geographies of older people—or children, or any other group— in isolation is not possible if we conceive of 'age as being produced in the interactions between different people' (Hopkins and Pain 2007). Such insights are fully in line with thinking about the relational nature of inequality, whereby the advantage experienced by one group can only be thought in the context of the disadvantage experienced by another (McCall 2005).

At the heart of an age-relationality framework are the concepts of the lifecourse, intergenerationality and intersectionality. A lifecourse approach recognises that the historical conditions and change experienced by ageing individuals over the course of their lives impact on their expectations and experiences in their present and future (Bengtson et al. 2005; Hörschelmann 2011; Barron 2019). Intergenerationality refers to the 'relations and interactions between generational groups' (Hopkins and Pain 2007). It assumes that people shape their age-related identities through interaction with others within the same or other age groups; therefore, children are always more-than children—and older people always more-than older people alone (Hopkins and Pain 2007). Finally, intersectionality refers to the intersection of gender, class, race, age and other social identities. In geography, the concept is used particularly in relation to their spatial specificity. The term is rooted in critical race theory and the analysis of the political movement of Black and other (ethnic) minority women in the United States; it was proposed in recognition of Black women's experiences of structural oppression and discrimination at the intersection of race and sex (Crenshaw 1989, 1991). It is an analytical tool that allows us to confront the complexity of specific phenomena (McCall 2005; Collins and Bilge 2016). As Hopkins and Pain (2007) note: 'the ways in which age is lived out and encountered are likely to vary according to different markers of social difference; the everyday experiences of people are diverse and heterogeneous'. Still, the intersectionality approach remains extremely rare in ageing studies (Calasanti and King 2015).

I should note that whilst the age-relationality framework appears very valuable in highlighting many aspects which should be included in any relational analysis of the ageing process, this framework does not readily lead us to consider other types of relations, for instance, interactions with the more-than-human (Whatmore 2002; Power 2008; Steele et al. 2019). Ageing is a process linked with a range of other processes, resources and conditions, including individual health and socioeconomic status, health care infrastructure and the condition of the local natural and built environments. However, approaches to the study of ageing at the intersection of multiple human and ecological systems are limited. A major hurdle to the integration of 'nature' or 'the built environment' in the social sciences continues to be the traditional divide between the natural and social sciences. This has led to disjointed advances in our understanding of the world: human behaviour is habitually excluded from biological models, which have focused mainly on the impact of human activity on biological systems, rather than human activity as part of biological systems. Similarly, although instances of the application of ecological principles to human activity exist—for example, the Chicago School, with 'community' as a key unit of analysis (e.g., Park et al. 1925)—'nature' has remained but a metaphor for researchers in the social sciences and humanities (Machlis et al. 1997; e.g., Kaika 2005; Heynen et al. 2006).

## DISCOVERING ECOSYSTEMS OF AGEING

As is probably evident from this introduction, my interest in researching and writing this book is a deeply personal one, and one that was not necessarily a natural fit with my professional training as an architect or scholarly expertise in urban studies. The experience of using a grounded theory approach (Glaser and Strauss 1967) in my previous work allowed me to attempt this project confidently, nonetheless. I was keen to discover a subject area on the ground, teasing out how older people in Bulgaria cope under conditions of permanent uncertainty. As such, I did not approach the project with preconceived hypotheses or questions. Rather, I designed the research to be specific enough to capture the diversity and complexity of older people's experiences—and broad enough to connect these to wider social, spatial, economic, ecological or other kinds of systems. The research included participant observation and in-depth interviews with older people in three neighbourhoods of Sofia, Bulgaria's capital and in the Village, located in the Bulgarian Balkans.

In order to enter the field as free as only possible of theoretical preconceptions related to the processes and experiences of ageing in a post-socialist society, I chose to use the human ecosystem (Machlis et al. 1997) as an underlying framework for the design of the interview guide and collection of data. I borrow the human ecosystem framework (HEF) (Machlis et al. 1997) from the field of ecology. It provides a framework for the consideration of different elements as part of larger human *and* ecological systems, thus facilitating research that acknowledges the co-existence and co-evolution of the human and non-human (or more-than human) from the onset.

In ecology, ecosystems are defined as networks of co-evolved organisms (Golley 1993). An ecosystem is 'any entity or natural unit that includes living and non-living parts interacting to produce a stable system in which the exchange of materials between the living and non-living parts follows circular paths' (Odum 1953). Human ecosystems are defined as coherent systems 'of biophysical and social factors capable of adaptation and sustainability over time' (Machlis et al. 1997). Their interlinked elements include natural, socioeconomic and cultural resources which are governed by social systems defined by institutions, cycles and order (see Fig. 1.1).

Notably, the HEF is primarily an auxiliary structure described by its component parts rather than by critical patterns or processes. Thus, although it suggests a comprehensive register of component parts to be

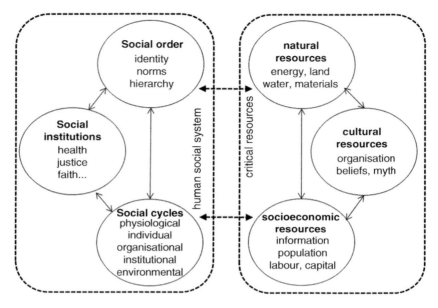

**Fig. 1.1**  The human ecosystem framework (based on Machlis et al. 1997)

considered in an enquiry, the HEF does not tell us just how these elements relate to each other, what kind of power relations are inscribed within these relationships or what kind of inequalities these may produce. To a certain extent, the HEF can therefore be regarded as politically neutral and value free. This is of advantages when attempting to understand, as I was, a subject area from the ground up. The following paragraphs provide a brief overview of the component parts of the HEF.

- Critical natural resources include energy, land, materials, nutrients and, of course, flora and fauna. For instance, *energy* is essential to the functioning of social systems and can vary by type, quality and flow (often subject to external control). *Land* is characterised by patterns of land cover type, ownership (public or private?), economic and cultural value and use. Land use change can impact on hierarchies of wealth, power and territory. *Materials* include lumber, glass, concrete and so on. The range of materials which flow within human ecosystems depends on consumption patterns, local culture and the level of economic development. The nature and norms for the use of materials can change with shifts in material flows (Machlis et al. 1997).

- Critical socioeconomic resources that are part of human ecosystems include, among others, information, population, labour and capital. At the heart of systems theory is *information* as a socioeconomic resource and how it flows through a system (von Bertalanffy 1968). Information can be transmitted, for instance, through genes, language, digital data or media. Both critical resources and the components of social systems depend on information flows. *Population* is a resource that provides labour, accumulates knowledge and consumes resources. *Capital* refers to the economic instruments for the production of commodities. Capital includes financial resources (money), technological tools and resource values (Machlis et al. 1997).

- Critical cultural resources include organisation, beliefs and myths. *Organisation* creates and sustains human social systems. Different cultures employ their organising skills differently to manage their resources. *Beliefs* are statements about reality—based on personal observation, mass media, faith, science and so on—that are accepted as true (Machlis et al. 1997).

- Social institutions in the human ecosystem include those aimed at organising health, justice, faith, commerce, education, government and sustenance. They can include households, community groups, government agencies and similar formations. For instance, *health institutions* include organisations and activities related to the health needs of the human ecosystem in question. They include primary care (serving health maintenance, such as general practitioners), secondary care (specialist care) and tertiary care (hospitalisation). Health institutions are measured by capacity (such as doctors per 1000 population) or outcomes (such as infant mortality rates). *Leisure* refers to the 'culturally influenced ways [in which] we use our nonwork time' (Machlis et al. 1997). The category includes formally, semi-formally and informally managed activities. *Government* is the result of the need for decision making across a variety of scales and includes the interaction between and processes of decision making within different political units. It relies, at least in theory, on the participation of citizens. Government has close links with and control of critical natural (and other) resources and can therefore have great impact on the whole human ecosystem (Machlis et al. 1997).

- Social cycles include physiological, individual, institutional and environmental cycles. *Physiological cycles* are cycles such as night and day, menstrual cycles or life cycles. They create patterns that allow to predict behaviours (e.g., energy demands during the day, weddings and funerals). Although relatively rigid, they impact human ecosystems at different scales. *Individual cycles* are personal: graveyard shifts, seasonal work or patterns of recreational activities. The use of natural resources and social institutions depends on individual cycles. Finally, *environmental cycles* are not socially constructed but have significant influence on human ecosystems in triggering ecosystem and social system responses (Machlis et al. 1997).
- Social order is constituted by the elements of identity, norms and hierarchy. *Identity* can be acquired during the lifecourse (e.g., political identity) or ascriptive (i.e., assigned by society from birth—e.g., caste or race). Human ecosystems are significantly influenced by age as an identity marker in that it impacts occupation and recreation activities, among others. Changes in identity may lead to changes in social norms. *Social norms* are formal and informal rules for behaviour, enforced by laws or monitored through the (dis)approval of social groups or communities, respectively. Formal and informal norms can stand in conflict to each other. They are linked with social institutions (for their enforcement) and the use of resources. Finally, *hierarchy* is omnipresent in social systems as a result of social differentiation (Machlis et al. 1997).

Because of the time and resource limitations to the research presented in this book, my adaptation of Machlis et al.'s (1997) HEF did not, by far, embrace all component parts included in the original HEF. I did, however, make every effort to address each of the main categories and as many component parts as realistically feasible as I developed the research design and instruments. Admittedly, this required an initial selection of categories (and component parts) that I considered most relevant to the subject area of the research, thus setting the boundaries to the project in a way that may have excluded important aspects of ageing in Bulgaria from the enquiry. I then constructed a semi-structured interview guide which I tested in pilots and then revised and adjusted in response to particularities I encountered. This included adjusting the language, for instance, avoiding the use of certain terms that caused confusion or simplifying questions so that they were as open-ended as possible to trigger detailed responses and narratives from interview participants.

The resulting interview guide comprised of 25 open-ended questions of varying complexity, ranging from 'which social class do you belong to' to 'how do you spend a typical day'. Questions covered all domains of the HEF: natural resources (energy and land), socioeconomic resources (information and population), cultural resources (organisation, beliefs), social institutions (health and capital), social cycles (individual across different temporal scales: daily, annual and lifecourse), and social order (identity, social norms and hierarchy). In this way, I hoped to cover aspects entangled in the ageing processes that would not necessarily be spoken about otherwise. Questions were asked in a way that allowed participants to respond in as much or as little detail as they desired.

Research took place between November 2015 and December 2017, with the majority of interviews completed between June and October 2016. I conducted half of the interviews myself; the other half were completed by a qualified research assistant. Initially, research participants were approached through social workers in local pensioners' clubs (in the city) and through local community leaders in the Village. Further participants were recruited through snowballing. We taped interviews with the agreement of participants. In some cases, participants were cautious of giving permission for the audio-recording of interviews, which is likely to be attributed to their fears of being reported—a shared experience during communism (see also Conkova et al. [2019]). Where this was the case, detailed notes were taken.

Interviews lasted between 30 minutes and two hours, reflecting the varying willingness of participants to share their views and experiences. They took place in community centres or the participant's home, according to their preference. Occasionally, they took place in local cafes or restaurants. Participants were interviewed alone or in the presence of a close relative or carer, which means that sometimes, responses did reflect ingroup effects. However, we took care during the interviewing process to probe for inconsistencies and clarify with the interviewees specific aspects in situations where more privacy was given. The interviews were accompanied by prolonged periods of observation, where I stayed around or participated in collective activities, such as singing, having lunch or coffee, or going for a walk.

In the process of transcription I translated all interviews and moved on to coding and analysis. Participant names were exchanged for pseudonyms to preserve the anonymity of participants. The analysis builds on deductive (theory-driven), inductive (data-driven) and in vivo (using participants' own words) codes. The analytical framework for the exploration of interview transcripts was developed from a thorough initial reading and the emerging themes, roughly following a grounded theory approach of constant comparative analysis (Glaser and Strauss 1967). As would be expected using this approach, the back-and-forth between my data and the extant literature led to the loss of the original HEF categories (or component parts) through which data was collected using the interview guide. This happened in the process of analysis and writing up—in favour of capturing emerging concepts and how they relate to one another, thus developing the theoretical interpretation of conditions on the ground as presented in this book.

My intention was to understand how older people in Bulgaria cope under conditions of all-encompassing cultural, socioeconomic and environmental transitions. Overall, we interviewed 74 people aged between 61 and 94 at the time of their interviews. Of these, 20 were located in each of the three selected neighbourhoods in Sofia and the remaining 14 in the Village. Whilst I was eager to capture the views and experiences of men and women and people from different ethnic backgrounds, the make-up of participants does not reflect this ambition. The vast majority in this study are women (60) and almost all are ethnically Bulgarian. A limitation of this research is therefore that, although concerned with widening inequalities in Bulgaria, it does not reflect sufficiently on inequalities experienced on the basis of ethnicity or religion. This leaves questions open for future research.

Finally, I should state that this is not a book that seeks to fit neatly into a specific academic field. True to the grounded theory approach, my research involved going back and forth between data collection and analysis, between the field and the literature. Despite having worked across multiple disciplinary domains and thematic fields in the past, of course this project required becoming familiar with unfamiliar bodies of literature, including a range of theories of ageing which I present in the previous section as well as throughout the rest of the book. I draw from a range of academic disciplines, including sociology, architecture, geography and environmental psychology.

I note here my position as a researcher in the study, as 'to ignore the role of the researcher [...] would be an oversight and would present a disingenuous view of the research' (Hall 2019). As I describe in the opening paragraphs to this book, I spent my childhood between Sofia and the Village. My own life is therefore intimately entangled with the places—and some of the people—whose lives I discuss in this book; therefore, I do not attempt to present the research as removed from this reality. This is in line with the Glaserian dictum of 'all is data' in grounded theory (Glaser 2002). The reader will encounter personal accounts, including retrospective autoethnography in places, which I have included to embrace (rather than conceal) my positionality. Indeed, the insights presented here are informed by my own experience of a 'translocal' and 'transnational' life and oscillating between the East and West. Many research participants have known me since I was a child, a part of their community. I returned after three decades of little or no contact, holding a PhD, living and working in a Western European country and—during significant periods of my fieldwork—being heavily pregnant or carrying in my arms a charming little baby. The combination of these factors seemed to put me in a position of advantage, allowing participants to feel at ease and to open up more readily than they may have, otherwise.

## Locating Ageing in the City and in the Village

The research presented in this book explores ageing in the context of rapidly transforming urban and rural Bulgaria through four case study areas: a town so small that throughout this book I call it 'the Village' to protect the anonymity of its residents, and three neighbourhoods in Sofia, the capital city. The case studies were selected to reflect on the differences between ageing in the Village and the city, taking into account the socioeconomic gradient between the three neighbourhoods in Sofia. As previously noted, rural Bulgaria is continuously depopulating, losing its working age population to cities and suffering from the consequences of rapid population ageing. Simultaneously, Sofia is drawing people from the countryside and from other cities, consequently becoming one of only few growing cities in Bulgaria; see Fig. 1.2). I provide here a brief overview of the selected case study areas, noting that they will come to the fore more forcefully in the chapters to follow.

The Village is formally a small town located on the southern slopes of Bulgaria's Stara Planina (see Fig. 1.3). The settlement became a town only in the late 1970s but in essence can be characterised as a typical Bulgarian village. The onset of the Transition in 1989 and the associated systemic changes, including substantial cuts to public funding, affected core

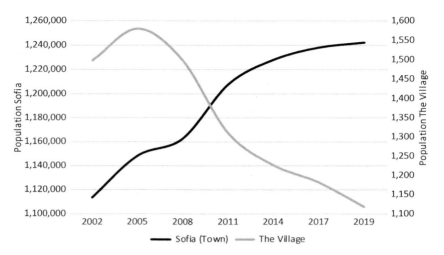

**Fig. 1.2** Population dynamics Sofia (Town) and the Village, 2002–2019. (Source: NSI 2020)

functions in the Village. A factory with headquarters in the nearby town, one of the very few still open in the region, continues to be the main employer of working age people from the Village to this day. There was a short period of collaboration with a corporation from Japan, resulting in the curious, yet short-lived, appearance of Japanese pensioners in the Village. They rented houses, learned the language and engaged with the villagers in the local community centre, once at the heart of the village community. Yet activities there came to a halt, eventually, and with no funds to maintain the building—a grand building constructed in the 1930—the community centre began to crumble. Thugs slowly took the furniture, the light fixtures, the electricity wires, the tiles, the ornaments, and the glass from its windows. The local furnace suffered a similar fate. Once famous in the region, it drew apprentices from all over to learn how to make bread. Today, the Village's sole donkey can often be found tied to the entrance to the small disintegrating building, the former bakery counter still standing its ground behind the broken facade. Today, the Village has some 1100 residents, women in the majority (NSI 2017). It is a rapidly declining population; the town lost 25% of its residents in only 11 years (Fig. 1.2). According to the statistics, there are some 50–60

**Fig. 1.3** A typical house in the Village. (Photograph: Iossifova, 2016)

deaths every year. People know when someone dies when the church bells ring. Twice for a woman. Thrice for a man.

Sofia's history dates back some 2000 years. Following Bulgaria's independence from the Ottoman Empire, the city grew from 12,000 residents in 1879 to some 300,000 in the 1930s; neighbourhoods east and south of the city core became fashionable and attracted affluent residents. Workers, rural-to-urban migrants and refugees were confined to neighbourhoods in the north and west (Brade et al. 2009).

Once the communists took over in 1944, Sofia expanded rapidly to reflect the development of the nation—Bulgaria was industrialising and Sofia became a magnet for factory workers and government officials. The capital city contained a historic central core hosting government, finance, management and leisure for the population, far beyond the city's boundaries. This core was surrounded by the neighbourhoods of the pre-war period, which, in turn, were enveloped by a ring of socialist-era developments of prefabricated concrete panel blocks. Such developments usually had their own shopping, medical, educational and other services, modelled after Soviet-style *microraions*. Industrial zones were located on the margins of the city, and Sofia continues to benefit from the large swaths of green created during this period (Staddon and Mollov 2000).

With the fall of communism in 1989, urban patterns began to change. For instance, the number of retail spaces in the city grew four times between 1990 and 2005, and much retail activity could be attributed to small private firms (including the cafe in the garage or the shop in the entry hall) (Hirt 2006). However, although the city centre remained a retail location, it became increasingly occupied by big-name Western brands; overall, retail expansion took place away from the centre. The retailing zones once planned by communist urban planners are now occupied by Western supermarket chains like Kaufland or Lidl (Staddon and Mollov 2000).

The legacy of the strategies promoted in the early 1990s by the World Bank and other international actors—including neoliberal policies around deregulation, decentralisation, privatisation and commodification—has led to the emergence of 'New Wealth and New Poverty' (Staddon and Mollov 2000). Private trade and services have grown based on the restitution of property to original owners and heirs and the sell-off of state assets (Vesselinov and Logan 2005). New construction is much smaller in scale and has little in common with the large residential complexes of the past. This is due to the withdrawal of the state from the production of housing (Hirt 2006). Bulgaria's entry into the EU has helped to blur its capital's defining post-socialist characteristics even further (Hirt 2006). Gated communities have become a symbol of Sofia's urban transition and expansion (Smigiel 2013). Income is now the most significant factor in residential choice.

Today, Sofia has areas of relative affluence and areas of relative deprivation as a result of uneven development. The selected case study areas reflect this diversity. They include Lozenets, an affluent inner suburb in the south-west once largely reserved for government and party officials (Staddon and Mollov 2000); Vazrazhdane, a central district hugely transformed during the 1970s and 1980s as a result of clearance and the construction of 'modern' higher-rise buildings; and Zapaden Park, a neighbourhood in the west characterised by the post-war socialist housing project constructed to house the capital's workers and other low-income population groups (see Fig. 1.4). The socioeconomic gradient described above is further reflected in the ethnic make-up of the case study areas,

**Fig. 1.4** Typical 1940s housing block in Zapaden Park, Sofia. (Photo: Iossifova, January 2016)

**Table 1.1** Ethnic composition of study areas in 2011, in percent

|  | Sofia (Mun) | Krasna Polyana | Vazrazhdane | Lozenets | The village |
|---|---|---|---|---|---|
| Bulgarian | 87.9 | 73.2 | 84.6 | 90.6 | 85.4 |
| Turkish | 0.5 | 0.4 | 0.5 | 0.6 | 0.0 |
| Roma | 1.5 | 14.6 | 1.0 | 0.1 | 0.5 |
| Other (incl. no answer) | 10.2 | 11.8 | 13.9 | 8.7 | 14.1 |

Source: Bespyatov (2020)

displaying an increase of ethnically Bulgarian population ratios from west (Krasna Polyana, where Zapaden Park is located) to east (Lozenets), going hand-in-hand with the decrease in the numbers of Roma (Table 1.1 and Fig. 1.5).

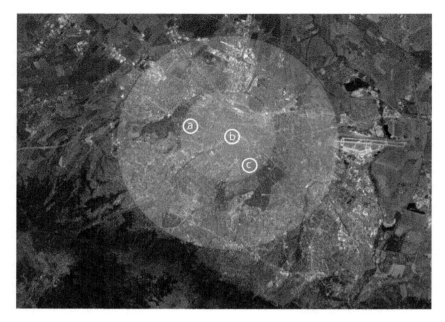

**Fig. 1.5** Aerial photograph of Sofia, showing the locations of the case study areas: (a) Zapaden Park; (b) Vazrazhdane; and (c) Lozenets. The concentric rings show the expansion of the city before 1944 (darker) and during state socialism (lighter). Aerial photograph derived from Google Maps in August 2020, alteration by Iossifova. (Imagery ©2020 TerraMetrics, Map data ©2020)

In this context, it should be noted that research on cities beyond the West (Edensor and Jayne 2011; Robinson 2006; Roy 2009) continues to mean research in and on the Global South, the former Third World, even as urbanism and urban studies are waking up to calls for more inclusive urban scholarship. Recent calls for the provincialising of global urbanism largely exclude the (post-socialist) Global East (Sheppard et al. 2013; Leitner and Sheppard 2016; Schindler 2017; Bhan 2019). Indeed, as Müller (2021) highlights, the cities explicitly named in recent manifestos for 'urbanism from the South' are almost exclusively cities associated with European colonialism. Although they are different in many ways, cities of the Global East continue to be interpreted and portrayed as slightly modified, possibly even modernised, versions of the Western capitalist city. However, they cannot be conflated with the city of the Global North (i.e., the neoliberal city) (Pacione 2001); nor can they be thought of as instantiations of the marginalised city of the Global South (Robinson 2002). Careful, nuanced scholarship of post-socialist urbanism, such as Hirt's (2012) on emerging post-socialist urban form, is scarce.

Finally, a note on the research settings: in some sense, my approach, albeit not ethnographic per se, is close to Hall's (2019) 'doorstep ethnographies', conducted near one's place of residence and work. I chose to work in locations that I knew in intimate ways because I spent most of my childhood and short periods of my adulthood there (Sofia's Zapaden Park and the Village). Of course, temporal and spatial distances have played their part to dissociate me from the close-knit communities with and within which the research took place. Regardless, those were deeply familiar settings that frequently brought back my own memories of people, places and experiences in the past. In order to acknowledge some of the associated emotional and cognitive bias, I therefore added additional sites to which I had no prior connection, including Bulgaria's former industrial centre Gabrovo—a rapidly depopulating and ageing town in the Balkans. Ultimately, findings in those settings did not add to those emerging from Sofia and the Village, thus I do not include them in this book.

## STRUCTURE OF THE BOOK

In this chapter, I have provided an overview of the challenges facing Bulgaria and its ageing population today. I presented a brief overview of current thinking around ageing in human geography and environmental gerontology in the lead-up to my analytical framework, the human ecosystem approach. The previous sections provided a detailed account of my methodological approach to unravelling entanglements of ageing and introduced the geographical context of the research.

In Chap. 2, I move to borrow from the lifecourse approach and its imperatives to link the lifecourse narratives of older people in Sofia and the Village with the wider socio-political context within which they occurred. I focus on the conceptual and empirical domains of migration, housing and labour (including care) and the two eras of communism (1944–1989) and the Transition (since 1990) to demonstrate how the personal and the circumstantial are entangled with the experience of ageing in Bulgaria. I argue that Bulgaria's systemic socio-political transitions over the last 80 or so years have shown enormous impact on people's lifecourses, including the ways in which older people position themselves and act in society today. Older people today have had to transition from the acceptance of their inability to change the course of their lives in major ways to navigating an emerging world in which the former promise of the state to take care of everyone and everything no longer held.

In Chap. 3, I detail how this newly found agency enables older people today to manage uncertainty in their everyday. I explore in detail their patterns of everyday life in the city and in the Village before turning to present their accounts of managing meagre pensions in taking on post-retirement jobs or keeping detailed track of their expenses; of sustaining ailing bodies when they have to rely on an increasingly unjust health system; and of making do in tattered homes, unable to heat them during the cold winter months. I point out that older people in the city and in the Village—or in cities and villages more generally—rely on vastly different resources and infrastructures. In closing, I argue that older people in Bulgaria, whether in cities or in villages, display astonishing resilience in the face of adverse financial, material, physical, political, social and cultural circumstances.

In Chap. 4, I turn to the notion and agenda of 'ageing in place' to argue that it is inherently inapt to capture or address the main concerns of older people in Bulgaria. The concept seems to assume that people age in static environments that are hardly subject to change—a notion which I reject, presenting how older people navigate and experience the transformation of their neighbourhoods in the city and the Village. I show how they move closer in the Village, drawing in on their established close-knit network of family, neighbours and friends and reaffirming their political and other value, norm and belief systems. I then argue that Bulgaria's rural-urban reciprocity has fostered profoundly translocal ageing processes, whereby older people oscillate between homes, social networks and environments in the city and in the Village. Finally, I show that recent trends in migration and processes of ageing are enmeshed, producing a kind of transnational ageing utterly incompatible with 'ageing in place' agendas. I argue that translocal ageing is both a condition and a coping mechanism, enabling older people today to cope with the reality of being both abandoned by the state and left behind by adult children and grandchildren.

I close this book with Chap. 5, arguing that older people in Bulgaria are indeed abandoned by the state; they have, however, over the course of their lifetimes developed adaptation strategies and coping mechanisms which allow them to act resiliently in the face of deteriorating financial, material, physical, political, social and cultural circumstances. I call for a deeper engagement with the Global East at the intersection of ageing studies and urbanism. I revisit the methodological approach and larger conceptual framework to suggest that understanding ageing processes through the human ecosystem lens creates insights that reverberate with recent agendas and approaches in relational geography.

## References

Andrews, G. J., Cutchin, M., McCracken, K., Phillips, D. R., & Wiles, J. (2007). Geographical gerontology: The constitution of a discipline. *Social Science & Medicine, 65*(1), 151–168.

Andrews, G. J., Milligan, C., Phillips, D. R., & Skinner, M. W. (2009). Geographical gerontology: Mapping a disciplinary intersection. *Geography Compass, 3*(5), 1641–1659.

Barron, A. (2019). More-than-representational approaches to the life-course. *Social & Cultural Geography*, 1–24. https://doi.org/10.1080/1464936 5.2019.1610486.

Bastia, T., Valenzuela, C. C., & Pozo, M. E. (2020). The consequences of migration for the migrants' parents in Bolivia. *Global Networks*. https://doi.org/10.1111/glob.12276.

Bengtson, V. L., Elder, G. H., & Putney, N. M. (2005). The Lifecourse perspective on ageing: Linked lives, timing, and history. In M. L. Johnson (Ed.), *The Cambridge handbook of age and ageing* (Cambridge handbooks in psychology) (pp. 493–501). Cambridge: Cambridge University Press.

Bespyatov, T. (2020). *Population statistics of Eastern Europe and former USSR*. Retrieved June 20, 2020, from http://pop-stat.mashke.org/bulgaria-ethnic-loc2011.htm.

Bhan, G. (2019). Notes on a southern urban practice. *Environment and Urbanization, 31*(2), 639–654. https://doi.org/10.1177/0956247818815792.

Brade, I., Smigiel, C., & Kovács, Z. (2009). Suburban residential development in post-socialist urban regions: The case of Moscow, Sofia, and Budapest. In H. Kilper (Ed.), *German annual of spatial research and policy 2009* (German annual of spatial research and policy) (pp. 79–104). Berlin, Heidelberg: Springer.

Buffel, T., & Phillipson, C. (2011). Experiences of place among older migrants living in inner-city neighbourhoods in Belgium and England. *Diversité urbaine, 11*(1), 13–37.

Buffel, T., & Phillipson, C. (2016). Can global cities be 'age-friendly cities'? Urban development and ageing populations. *Cities, 55*, 94–100. https://doi.org/10.1016/j.cities.2016.03.016.

Buffel, T., Phillipson, C., & Scharf, T. (2012). Ageing in urban environments: Developing 'age-friendly' cities. *Critical Social Policy, 32*(4), 597–617. https://doi.org/10.1177/0261018311430457.

Buffel, T., Phillipson, C., & Scharf, T. (2013). Experiences of neighbourhood exclusion and inclusion among older people living in deprived inner-city areas in Belgium and England. *Ageing & Society, 33*(Special Issue 01), 89–109. https://doi.org/10.1017/S0144686X12000542.

Calasanti, T., & King, N. (2015). Intersectionality and age. In J. Twigg & W. Martin (Eds.), *Routledge handbook of cultural gerontology* (pp. 193–200). London and New York: Routledge.

Central Intelligence Agency. (2017). *The world factbook.* Retrieved August 15, 2017, from https://www.cia.gov/library/publications/the-world-factbook/geos/bu.html.

Chłoń-Domińczak, A., Kotowska, I. E., Kurkiewicz, J., Abramowska-Kmon, A., & Stonawski, M. (2014). *Population ageing in Europe: Facts, implications and policies.* Brussels: European Commission.

Cloutier-Fisher, D., & Joseph, A. E. (2000). Long-term care restructuring in rural Ontario: Retrieving community service user and provider narratives. *Social Science & Medicine, 50*(7–8), 1037–1045.

Collins, P. H., & Bilge, S. (2016). *Intersectionality.* Cambridge: Polity Press.

Conkova, N., Vullnetari, J., King, R., & Fokkema, T. (2019). "Left like stones in the middle of the road": Narratives of aging alone and coping strategies in rural Albania and Bulgaria. *The Journals of Gerontology: Series B, 74*(8), 1492–1500.

Crampton, R. J. (2005). *A concise history of Bulgaria.* Cambridge: Cambridge University Press.

Crenshaw, K. (1989). Demarginalizing the intersection of race and sex: Black feminist critique of antidiscrimination doctrine, feminist theory and antiracist politics. *University of Chicago Legal Forum, 1989,* 139–168.

Crenshaw, K. (1991). Mapping the margins: Intersectionality, identity politics, and violence against women of color. *Stanford Law Review, 43*(6), 1241–1300.

Cutchin, M. P. (2009). *Geographical gerontology: New contributions and spaces for development.* Oxford: Oxford University Press.

Davis, F. (1979). *Yearning for yesterday: A sociology of nostalgia.* New York: Free Press.

Edensor, T., & Jayne, M. (Eds.). (2011). *Urban theory beyond the west: A world of cities.* London and New York: Routledge.

Estes, C. L. (2004). Social security privatization and older women: A feminist political economy perspective. *Journal of Aging Studies, 18*(1), 9–26. https://doi.org/10.1016/j.jaging.2003.09.003.

eurostat. (2020). *Income poverty statistics.* Brussels: eurostat.

Evans, M. (2013, 31 December 2013). Ghost towns left by Bulgarians seeking work in UK. The Telegraph. Retrieved from https://www.telegraph.co.uk/

news/uknews/immigration/10545458/Ghost-towns-left-by-Bulgarians-seeking-work-in-UK.html.

Gancheva, V., & Chengelova, E. (2006). Social Services for the Elderly [*Socialni Uslugi za Horata ot Tretata Vuzrast*]. Retrieved from http://www.omda.bg/public/biblioteka/vyara_gancheva/vyara_emi/vyara_emilia_00.htm.

Gardner, P. J. (2011). Natural neighborhood networks—Important social networks in the lives of older adults aging in place. *Journal of Aging Studies, 25*(3), 263–271.

Ghodsee, K. (2004). Red nostalgia? Communism, women's emancipation, and economic transformation in Bulgaria. *L'Homme: Zeitschrift für Feministische Geschichtswissenschaft, 15*(1), 23–36.

Glaser, B. G. (2002). Constructivist grounded theory? *Forum Qualitative Sozialforschung/Forum: Qualitative Social Research, 3*(3), Art. 12.

Glaser, B. G., & Strauss, A. L. (1967). *The discovery of grounded theory: Strategies for qualitative research.* New York: Aldine Transaction.

Golant, S. M. (2014). Age-friendly communities: Are we expecting too much? In *IRPP insight*. Montreal: Institute for Research on Public Policy.

Golley, F. B. (1993). *A history of the ecosystem concept in ecology.* New Haven, CT: Yale University Press.

Goody, J. (1996). Comparing family systems in Europe and Asia: Are there different sets of rules? *Population and Development Review, 22*(1), 1–20.

Greed, C. (2019). Join the queue: Including women's toilet needs in public space. *The Sociological Review, 67*(4), 908–926. https://doi.org/10.1177/0038026119854274.

Hall, S. M. (2019). *Everyday life in austerity: Family, friends and intimate relations.* Cham: Palgrave Macmillan.

Heynen, N. C., Kaika, M., & Swyngedouw, E. (Eds.). (2006). *In the nature of cities: Urban political ecology and the politics of urban metabolism.* London: Routledge.

Hirt, S. (2006). Post-socialist urban forms: Notes from Sofia. *Urban Geography, 27*(5), 464–488. https://doi.org/10.2747/0272-3638.27.5.464.

Hirt, S. (2012). *Iron curtains: Gates, suburbs and privatization of space in the post-socialist city.* Chichester: John Wiley & Sons.

Hodge, G. (2008). *The geography of aging: Preparing communities for the surge in seniors.* Montreal: McGill-Queen's Press-MQUP.

Hoff, A., & Perek-Bialas, J. (2015). Introduction: Towards a sociology of ageing in central and Eastern Europe. *Studia Socjologiczne, 2*, 9.

Hopkins, P., & Pain, R. (2007). Geographies of age: Thinking relationally. *Area, 39*(3), 287–294.

Hörschelmann, K. (2011). Theorising life transitions: Geographical perspectives. *Area, 43*(4), 378–383. https://doi.org/10.1111/j.1475-4762.2011.01056.x.

Iossifova, D. (2015). Borderland urbanism: Seeing between enclaves. *Urban Geography*, *36*(1), 90–108. https://doi.org/10.1080/0272363 8.2014.961365.

Kaika, M. (2005). *City of flows: Modernity, nature, and the city*. New York and London: Routledge.

King, R., & Lulle, A. (2016). Research on migration: Facing realities and maximising opportunities. In *A policy Review*. Luxembourg: Publications Office of the European Union.

King, R., Lulle, A., Sampaio, D., & Vullnetari, J. (2017). Unpacking the ageing–migration nexus and challenging the vulnerability trope. *Journal of Ethnic and Migration Studies*, *43*(2), 182–198.

Kulcsár, L. J., & Brădăţan, C. (2014). The greying periphery—Ageing and community development in rural Romania and Bulgaria. *Europe-Asia Studies*, *66*(5), 794–810. https://doi.org/10.1080/09668136.2014.886861.

Latour, B. (2005). *Reassembling the social: An introduction to actor-network-theory*. Oxford: Oxford University Press.

Leitner, H., & Sheppard, E. (2016). Provincializing critical urban theory: Extending the ecosystem of possibilities. *International Journal of Urban and Regional Research*, *40*(1), 228–235.

Lorimer, H. (2005). Cultural geography: The busyness of being 'more-than-representational'. *Progress in Human Geography*, *29*(1), 83–94. https://doi.org/10.1191/0309132505ph531pr.

Machlis, G. E., Force, J. E., & Burch, W. R. (1997). The human ecosystem part I: The human ecosystem as an organizing concept in ecosystem management. *Society & Natural Resources: An International Journal*, *10*(4), 347–367.

Manthorpe, J., Malin, N., & Stubbs, H. (2004). Older people's views on rural life: A study of three villages. *Journal of Clinical Nursing*, *13*, 97–104.

McCall, L. (2005). The complexity of Intersectionality. *Signs: Journal of Women in Culture and Society*, *30*(3), 1771–1800. https://doi.org/10.1086/426800.

Müller, M. (2020). In search of the global east: Thinking between north and south. *Geopolitics*, *25*(3), 734–755. https://doi.org/10.1080/1465004 5.2018.1477757.

Müller, M. (2021). Footnote urbanism: The missing east in (not so) global urbanism. In M. Lancione & C. McFarlane (Eds.), *Thinking global urbanism: Essays on the City and its future*. London: Routledge.

Nordbakke, S., & Schwanen, T. (2014). Well-being and mobility: A theoretical framework and literature review focusing on older people. *Mobilities*, *9*(1), 104–129. https://doi.org/10.1080/17450101.2013.784542.

NSI. (2020). *INFOSTAT*. Retrieved June 20, 2020, from https://infostat.nsi.bg/infostat/.

NSI. (2017). National Statistical Institute. Retrieved from http://www.nsi.bg/en.

Odum, E. P. (1953). *Fundamentals of ecology*. Philadelphia: Saunders.

Oswalt, P. (2005). Shrinking cities: International research. *Ostfildern-Ruit: Hatje Cantz, 1.*

Pacione, M. (2001). *Urban geography: A global perspective.* Oxon: Routledge.

Pallagst, K., Aber, J., Audirac, I., Cunningham-Sabot, E., Fol, S., Martinez-Fernandez, C., et al. (Eds.). (2009). *The future of shrinking cities: Problems, patterns and strategies of urban transformation in a global context.* Berkeley, CA: Center for Global Metropolitan Studies, Institute of Urban and Regional Development, and the Shrinking Cities International Research Network.

Park, R. E., Burgess, E. W., & McKenzie, R. D. (1925). *The City: Suggestions for the study of human nature in the urban environment.* Chicago: University of Chicago Press.

Peace, S., Kellaher, L., & Holland, C. (2005). *Environment and identity in later life.* London: McGraw-Hill Education.

Phillips, D. R., Siu, O. L., Yeh, A. G. O., & Cheng, K. H. C. (2004). Ageing and the urban environment. In *Ageing and place: Perspectives, policy, practice* (pp. 147–163). New York: Routledge Taylor & Francis Group.

Plyushteva, A., & Schwanen, T. (2018). Care-related journeys over the life course: Thinking mobility biographies with gender, care and the household. *Geoforum, 97,* 131–141. https://doi.org/10.1016/j.geoforum.2018.10.025.

Power, E. (2008). Furry families: Making a human–dog family through home. *Social & Cultural Geography, 9*(5), 535–555. https://doi.org/10.1080/14649360802217790.

Ratti, C., & Townsend, A. (2011). The social nexus. *Scientific American, 305*(3), 42–49.

Robinson, J. (2002). Global and world cities: A view from off the map. *International Journal of Urban and Regional Research, 26*(3), 531–554. https://doi.org/10.1111/1468-2427.00397.

Robinson, J. (2006). *Ordinary cities: Between modernity and development.* London: Routledge.

Roy, A. (2009). The 21st-century metropolis: New geographies of theory. *Regional Studies, 43*(6), 819–830.

Scharf, T. (2003). *Older people living in deprived Neighbourhoods: Social exclusion and quality of life in old age.* Swindon: ESRC.

Scharf, T., Phillipson, C., & Smith, A. (2005). Social exclusion of older people in deprived urban communities of England. *European Journal of Ageing, 2*(2), 76–87. https://doi.org/10.1007/s10433-005-0025-6.

Schindler, S. (2017). Towards a paradigm of southern urbanism. *City, 21*(1), 47–64. https://doi.org/10.1080/13604813.2016.1263494.

Schwanen, T., & Páez, A. (2010). The mobility of older people—An introduction. *Journal of Transport Geography, 18*(5), 591–595. https://doi.org/10.1016/j.jtrangeo.2010.06.001.

Schwanen, T., & Ziegler, F. (2011). Wellbeing, independence and mobility: An introduction. *Ageing and Society, 31*(5), 719–733. https://doi.org/10.1017/S0144686X10001467.

Schwanen, T., Banister, D., & Bowling, A. (2012a). Independence and mobility in later life. *Geoforum, 43*(6), 1313–1322. https://doi.org/10.1016/j.geoforum.2012.04.001.

Schwanen, T., Hardill, I., & Lucas, S. (2012b). Spatialities of ageing: The co-construction and co-evolution of old age and space. *Geoforum, 43*(6), 1291–1295. https://doi.org/10.1016/j.geoforum.2012.07.002.

Sheppard, E., Leitner, H., & Maringanti, A. (2013). Provincializing global urbanism: a manifesto. *Urban Geography, 34*(7), 893–900.

Skinner, M. W., Cloutier, D., & Andrews, G. J. (2015). Geographies of ageing: Progress and possibilities after two decades of change. *Progress in Human Geography, 39*(6), 776–799. https://doi.org/10.1177/0309132514558444.

Smigiel, C. (2013). The production of segregated urban landscapes: A critical analysis of gated communities in Sofia. *Cities, 35*, 125–135. https://doi.org/10.1016/j.cities.2013.06.008.

Smith, A. E. (2009a). Ageing in deprived neighbourhoods. In *Ageing in urban neighbourhoods* (Place attachment and social exclusion) (1st ed., pp. 85–134). Bristol: Bristol University Press.

Smith, A. E. (2009b). *Ageing in urban neighbourhoods: Place attachment and social exclusion.* Bristol: Policy Press.

Smith, A. E., Sim, J., Scharf, T., & Phillipson, C. (2004). Determinants of quality of life amongst older people in deprived neighbourhoods. *Ageing & Society, 24*, 793–814.

Staddon, C. (2004). The struggle for Djerman-Skakavitsa: Bulgaria's first post-1989 "water war". In *Drought in Bulgaria: A contemporary analog of climate change* (pp. 289–306). Aldershot: Ashgate Press.

Staddon, C., & Mollov, B. (2000). City profile: Sofia, Bulgaria. *Cities, 17*(5), 379–387. https://doi.org/10.1016/S0264-2751(00)00037-8.

Steele, W., Wiesel, I., & Maller, C. (2019). More-than-human cities: Where the wild things are. *Geoforum, 106*, 411–415. https://doi.org/10.1016/j.geoforum.2019.04.007.

The Royal Society. (2012). *People and the planet: The royal society science policy centre report 01/12* (S. Policy, Trans.). London: The Royal Society.

Thrift, N. (2007). *Non-representational theory: Space, politics, affect.* Oxon: Routledge.

Tsekov, N. (2017). "Authentic" rural depopulation in Bulgaria. *Naselenie (Население), 35*(2), 85–99.

Vesselinov, E., & Logan, J. R. (2005). Mixed success: Economic stability and urban inequality in Sofia. In F. I. Hamilton, K. D. Andrews, & N. Pichler-

Milanović (Eds.), *Transformation of cities in central and Eastern Europe: Towards globalization* (pp. 364–398). Tokyo: United Nations University Press.

von Bertalanffy, L. (1968). *General systems theory*. New York: Braziller.

Whatmore, S. (2002). *Hybrid geographies: Natures, cultures, spaces*. London: Sage.

WHO. (2007). *Global age-friendly cities: A guide*. Geneva: World Health Organization.

WHO. (2019, October 17). *Nursing and elderly home beds per 100 000*. European Health Information Gateway (Ed.).

World Bank. (2017). *Cities in Europe and Central Asia: Bulgaria*. Washington, DC: World Bank.

Zhecheva, A. (2007). *Handbook for the social Assistant in Bulgaria*. Sofia: Ministry of Labor and Social Policy.

Ziegler, F., & Schwanen, T. I. M. (2011). 'I like to go out to be energised by different people': An exploratory analysis of mobility and wellbeing in later life. *Ageing and Society*, *31*(5), 758–781. https://doi.org/10.1017/S0144686X10000498.

# CHAPTER 2

# Lives in Broad Strokes: Navigating Transitions, Disruptions and Uncertainty

**Abstract** Iossifova adopts the lifecourse approach and its imperatives to link the lifecourse narratives of older people in Bulgaria's capital Sofia and the Village in the Bulgarian Balkans with the wider socio-political context within which they occurred. She focuses on the conceptual and empirical domains of migration, housing and labour (including care) and the two eras of communism (1944–1989) and the Transition (since 1990) to demonstrate how the personal and the circumstantial are entangled with the experience of ageing in Bulgaria. She argues that older people today developed agency, having had to transition from the acceptance of their inability to change the course of their lives to navigating an emerging world in which the former promise of the state to take care of everyone and everything no longer held.

**Keywords** Bulgaria • Ageing population • Older people • Lifecourse • State socialism • Rural-urban reciprocity • Transition • Post-socialism • Migration • Housing • Care • Agency • Uncertainty

Once upon a time, there was a cruel king who thought that older people were of no use. He ordered that all older people in the kingdom be killed. One of his advisors, however, felt sorry for his old father and hid him in a backroom of his house to spare his life. Meanwhile, the summer brought an unseen drought that caused great damage and loss. Rivers and wells

© The Author(s) 2020
D. Iossifova, *Translocal Ageing in the Global East*,
https://doi.org/10.1007/978-3-030-60823-1_2

dried out, granaries were depleted and people were starving. No seed was left for sowing. The king ordered his advisors to find a solution within 24 hours or else he threatened to take their lives. When the advisor who had spared his father's life returned home, the old man listened to his anguish and offered advice on what to tell the king the following day: to order his peasants to dig through the kingdom's anthills. There, they would find plenty of seeds. And indeed, the peasants retrieved more than enough seeds from the anthills. The king was very pleased and wanted to know how his advisor came up with this recommendation. Once he admitted that his father was the source of advice, the king announced a new decree: never again should old people be harmed. Ever since, younger people show their respect in bowing and giving way when they encounter older people in the streets.

*    *    *

Older people invariably mention this tale in the course of conversations about their lives in the context of a transitioning society in Bulgaria. The tale is a reflection of a number of feelings and perceptions, among them the contradiction between the roles of older people in the past and their position in society today. In this chapter, I have two aims: to sketch how older people's lives have been shaped by the historical context within which they unfolded and to locate how and where older people position themselves within the broader framework of an existing (or perceived) social order today. I do this in an attempt to foreground just how the experiences of a shared past may have contributed to the positionality of older people today.

I build on interviews with older people in Sofia and in the Village, structured around the registers of the human ecosystem framework (Machlis et al. 1997, see Chap. 1). In particular, I work here through the accounts of older people in response to questions around *social cycles* (at the scale of the lifecourse) and *social order*, constituted by the elements of identity (ethnicity, religion, class, political beliefs); norms (expectations of the younger generation) and hierarchy (here, discussed through the lens of participation in decision making within the family and in society). I draw on Bengtson et al. (2005) and the five overlapping principles of a lifecourse perspective to interpret the interviews and construct the narratives presented here. These principles are: (a) lives are shaped by social and historical context; (b) the timing of individual transitions is relative to social context; (c) ageing is a life-long process, therefore what happened in the past has consequences for later life stages; (d) lives are interconnected and (e) individuals have agency to construct their lives.

These principles posit that the characteristics of a historical period shape how individuals think, feel and interact with others or how they view the world. As a result, the opportunities and constraints created by larger-scale conditions can shift the trajectories of individual lives in limiting choice and behavioural options. External events and transitions are experienced by and impact differently on individuals and their lifecourses depending on the stage of a person's life within which they occur. Of course, people live their lives embedded in relationships with others and they are influenced by them; as Hall (2019) eloquently puts it: 'it is difficult to extract individual stories and experiences because they are jumbled together in a knot of emotion, response and circumstance'. Lives are 'linked over time in relation to changing times, places, and social institutions' (Bengtson et al. 2005). Ageing is understood as a life-long process and therefore 'relationships, events, and behaviours of earlier life stages have consequences for later life relationships, statuses, and wellbeing' (Bengtson et al. 2005).

The lifecourse approach brings to the fore historical conditions and change experienced by the ageing individual over the course of their life. It acknowledges that 'we live dynamic and varied lifecourses which have, themselves, different situated meanings' (Hopkins and Pain 2007). The approach recognises that the experiences of people over time have impact on their expectations for their present and future. For instance, in order to understand why older people may be vulnerable, we may need to account for factors that influenced the course of their lives in the past (e.g., chronic disability) (Windle 2011).

In the following sections, I begin in linking the lifecourse narratives of older people in Sofia and in the Village with the wider socio-political context in Bulgaria within which they occurred. In the simplest of terms, participants categorised these stages as childhood, marriage, child rearing, labour and retirement, although oftentimes these did not arise chronologically; nor could they be readily disentangled where they occurred simultaneously (see also Barron 2019). Building on elements of the lifecourse approach, I use the two eras of state socialism (1944–1989) and the Transition (since 1990) and the conceptual domains of migration, housing and labour (including care) to demonstrate the entanglement of the personal and the circumstantial in the experience of ageing in Bulgaria. I then turn to argue that older people in Bulgaria today have had to find agency in adapting identities, roles and expectations to an emerging state of permanent uncertainty.

## FROM THE VILLAGE TO THE CITY: THE STATE TAKES CARE
## OF EVERYBODY AND EVERYTHING

Lyudmila was born in 1930 in a town approximately a three-hour drive from Sofia. Her parents were devoted members of the communist party and brought her up to be hardworking and realistic in her outlook. Aged 20, she moved to Sofia with her parents and lived with them whilst studying to become a doctor. Upon graduating from university in 1954, she got married and moved to a rental apartment where she had her first child that same year. In 1957, the young family moved into a modern apartment block in Vazrazhdane, close to the city centre. Lyudmila has been living there since, therefore sharing intimate relations with neighbours, their children and grandchildren.

Stoyanka was born in 1938 to a *Karakachan* (a minority of Greek origin) father and a Greek mother in the Village. She found a husband in the village and together, they moved to the industrial town of Gabrovo to work in a factory. They lived in an apartment provided by their employer and soon started a family. Environmental conditions in the Village were deemed better, so once old enough to go to nursery, the child was sent back to the Village to live with her grandparents. Stoyanka remembers fondly the close-knit community she was part of during her working life under state socialism, which included a whole range of recreational activities, such as hiking in the Balkans, birthday celebrations and collective cleaning of the urban environment.

Lyudmila and Stoyanka's trajectories are typical for a generation that not only witnessed the rapid urbanisation of the country in the course of a few decades, but consists largely of rural-to-urban migrants. Although cities in Bulgaria began to grow at the start of the twentieth century, Bulgaria's urban transformation took place within the three decades between 1950 and 1980, when the percentage of the country's urban population increased from 20 to 70% (Hoffman and Koleva 1993). From 1949, the country operated a planned economy, in the first years of state socialism placing the emphasis on the development of heavy industry (Ivanov 2007). The new industrial economy created jobs for migrants from the countryside who were prepared to go where they were needed

(Smollett 1989). Some older people today remember the early years after the communist takeover, when 'youth brigades' made up of youngsters from the countryside spent months building the country's much-needed infrastructure—roads, bridges, dams—in exchange for economic, technical and political education which they could then put to good use upon their return to their home village and (industrialised) agricultural work (Smollett 1989). Originally, workers in the city received subsistence support from their villages and kept returning to the village in order to support their ageing kin with agricultural work, consolidating, over time, a unique kind of rural-urban reciprocity that still exists today (Smollett 1985, 1989).

Urbanisation was fully in line with the state's vision of modernity and controlled through instruments such as the urban residence permit, the so-called *zhitelstvo*, required from 1947 in order to live in Sofia and from 1955 in cities across the nation (Konstantinov and Simić 2001). Such an internal movement control system existed in most socialist countries and is comparable to the Chinese *hukou*, a household registration system which designates rights to housing, education and health care, among others, based on place of birth (e.g., Chan and Zhang 1999). Older people report on the restrictions that such instruments imposed upon their life choices, even, as in the case of Bonislava, altering their respective career paths:

> I was born in [...] a village in the north of Bulgaria. I came to Sofia for work. I was not allowed to work in my profession at the beginning because back then a residence [permit] for Sofia was still required. I started as a worker in a factory. I then graduated, after studying in the evenings, and then I was able to work in planning. (Bonislava, 64, July 2016)

Resulting from massive rural-to-urban migration in the 1940s to 1960s, Bulgaria's cities—and Sofia in particular—began to struggle with undersupply of housing and worsening housing conditions, such as overcrowding. For political reasons, but also to make housing available quickly, a law came into operation in 1948 which effectively allowed the dispossession of Bulgaria's former political elite from any second dwellings; furthermore, they were forced to accommodate workers, rural-to-urban migrants and the new political elite in their primary houses or apartments (Parusheva and Marcheva 2010). Unlike most other countries in the Eastern Bloc, however, in addition to public (in theory owned by the people, in practice owned by the state) and cooperative ownership (mostly applied to agricultural land), Bulgaria retained a degree of private home ownership after WWII.

From 1958, once the urban housing crisis was in full bloom, the construction of housing became a national project implemented by state enterprises and, to a much smaller extent, by so-called housing cooperatives. Cooperatives were built by groups of friends or relatives. Small blocks across three floors, they were built to save money and space, containing apartments with two rooms or less organised around a central living room (Parusheva and Marcheva 2010):

> My husband was from Sofia [...] We used to live in a little house with the baby and his parents, but after his parents died [in 1962] we demolished the little house and built this building. [...] We got together with other people and built this block. The flat turned out a bit small, because we had to retract a bit to keep the necessary distance from the building next door. So it was a bit narrow, but we lived nicely. We have this room, another room, a kitchen, a bath. I am happy and satisfied. And at my age, it's more than enough. (Tanya, 76, June 2016)

Cooperatives had to pay much higher prices for goods and services, meaning that in reality, only the better-off could afford to join housing cooperatives. State enterprises were in a position of privilege as they had access to heavy state subsidies for materials, transport and land (Vesselinov 2004). Large state-owned enterprises collaborated with municipalities and together took on the role of investors and developers, acquiring land, contracting and financing construction and allocating units for sale or rent (Hoffman and Koleva 1993). New-built residential developments were designed with the aim to control and discipline a growing urban population. The state applied the Soviet model common to many Eastern Bloc countries (e.g., China's Maoist city; see Gaubatz 1998): urban areas were segregated according to functions and devised working, residential, recreational and service zones. New-built residential developments usually contained their own, carefully calculated provisions for services, ranging from shopping to education (Staddon and Mollov 2000).

In addition to regulated home ownership, the state had put in place distributive measures designed to cater to communist ideals, such as housing as a human right. So-called housing committees were in charge of allocating dwellings on a case-by-case basis. Housing was distributed according to need and regardless of income, benefiting 'young couples, large families, qualified specialists and workers living in bad conditions' (Vesselinov 2004). However, tenants were frequently allocated to buildings that were not purpose-built, such as warehouses or factory buildings (Parusheva and Marcheva 2010). Workers were not satisfied:

> They gave us a room in a warehouse, but without a toilet. So we had to go and use the public toilet. Finally, at the end, when they distributed apartments to the factory workers, they offered [my parents] an apartment, but for 6000BGN, which they had to collect from relatives and friends. (Blagun, 66, July 2016)

In this context, it should be noted that public and private relations in Bulgaria were governed by values that stand in stark contrast to 'the idealised Western model of a civil society based on universalistic principles' (Konstantinov and Simić 2001). Similar to China's institution of *guanxi* (e.g., Fei 1992), informal networks and reciprocal exchanges—*vraski* (connections)—facilitate much of society's business in Bulgaria. Controlled by the state, housing distribution resulted in its ability to reward a politico-military, industrial and intellectual elite with larger apartments in central locations (Vesselinov 2004; Parusheva and Marcheva 2010):

> They were seeking party members when I worked in the factory, so I asked my mother about it, and she said to become a member. Not that I am particularly leftist in my political views. But mother said: one day you may need something. It's good to be a member of the party then. (Blagun, 66, July 2016)

When it came to purchasing an apartment, even the better off and politically higher-ranking officials were dependent on the distribution system. Applications for housing loans were approved automatically once a household had obtained permission to buy or build a home from the municipality. Housing loans were provided by the State Savings Bank at an interest rate of 2% and to the maximum of 20,000 BGN over 25 years. However, the ownership of individual property was limited to one home in the city and one holiday home ('villa') per household (Hoffman and Koleva 1993; Vesselinov 2004). Such second homes were intended to compensate the rural-to-urban migrants of early socialist times for the restrictions and limitations of life in prefabricated complexes in the city (Yoveva et al. 2000). Villas became increasingly common in the 1970s, providing now established urban citizens with the legitimate opportunity to spend their weekends and holidays in the countryside. In this way, the 'villa' contributed to the consolidation of the initial 'pattern of rural-urban reciprocity' (Smollett 1985, 1989; Konstantinov and Simić 2001)— whereby rural-to-urban migrants kept going back and forth between the city and their village of origin in order to support ageing parents and kin in exchange for sustenance. Bulgaria's urban transformation remains partially incomplete at least in parts due to the peculiar kind of multi-local rural-urban existence enabled by the institution of the 'villa'.

Needless to say that the domains of housing, labour and caring overlapped and intersected where they depended on the state's distributional system coupled with the co-habitation of multiple generations due to the inability of married couples to afford their own apartments. In the 1950s, 1960s and 1970s, the lives of today's older people were shaped by caring responsibilities towards young children, for which this generation received support from a well-meaning state that provided free and secure childcare—but also from co-resident parents and grandparents. That said, the struggle for women's equality was an ideological commitment made by the communist state from the beginning; as a consequence gender relations in Bulgaria became profoundly remade (Ghodsee 2004).

The traditional and patriarchal social structure (including the extended family unit *zadruga*, comprising 10–20 blood-related small families who lived and worked together) had been disintegrating ever since the 1850s (Federal Research Division 2017). Women had been eking out equal rights, such as the right to inherit property, earning Bulgaria the reputation of being home to 'one of the best developed women's movements in Europe' in the 1920s (Ghodsee 2004). Subsequently, like Lyudmila and Stoyanka, women under state socialism participated in the labour force, strongly supported by policies that allowed them to achieve a good balance between paid employment and family responsibilities. Such policies included the creation of public canteens and child care facilities, maternity leaves (paid leave started 45 days before delivery and continued until the child's second birthday, with the option to extend this by another year of unpaid leave), child allowances and early retirement at age 55.

Some argue that the generous provision of public services did, in fact, create a kind of 'socialist paternalism' (Verdery 1996) in installing dependency on the state for men and women. The traditional role of the man as the provider and head of the household, in the past expected to make all decisions for younger and, in particular, female members of the family, had been taken over by the benevolent state in exchange for the familial loyalty it expected from its people. Others interpret the state's encouragement of women to participate to equal parts in family, work and community life as the 'triple burden' created under state socialism, which explains why many women experienced its fall as liberation (Einhorn 2002). However, this interpretation is not echoed in the narratives of the older women who participated in this research. Most felt supported in many ways under the 'old system' (socialism), by both the state and older generations. Their parents and even grandparents, just like in Stoyanka's case, often took on the role of the main childcare provider so that women could participate in the workforce and build careers.

> I trained as an architect. My father worked for the [government], so I started work for the government. [...] Back in the days they sent us to the countryside, to the villages, after graduating from university. We built houses for shepherds and sheds for sheep, stuff like that. I used to travel a lot throughout Bulgaria. When I was away, my husband took care of the kids, with the help of my mother. (Boyana, 67, June 2016)

Slightly later in life, these women would return the favour when their parents and in-laws became older and needed care. Still in 1980, some 17% of all households in Bulgaria span three or more generations (Federal Research Division 2017). Today's older women often found themselves cohabiting with and looking after elderly parents in ill health.

Lastly, although individual agency and choice were restricted under state socialism, jobs were secure. Men and women of working age generally enjoyed the security that state-guaranteed jobs offered (Smollett 1989). Their generation entered the workplace and built careers in a secure environment. Although possibly not fully satisfied with their work environment, where they were allocated jobs or who they had to work with, they could rest assured that upon graduation from school or university, there would be a position waiting for them. Of course, just as with the allocation of housing and all other benefits, better jobs were linked with political alignment:

> Political views were necessary when you wanted to get a job and were afraid that they will give it to someone else. I have worked in the dirtiest places where no one wanted to go, so I didn't have to hold political views. (Yana, 74, July 2016)

During Bulgaria's 45 years under socialism, the state, as Ghodsee (2004) writes, 'guaranteed economic security and the means for satisfying basic human needs'. It claimed that it 'took care of everybody and everything' whilst more or less surreptitiously controlling all spheres of public and private (Parusheva and Marcheva 2010). All this came to a grinding halt with the onset of the Transition in the last moments of the 1980s.

## THE LONG MARCH TO FREEDOM: ADJUSTING TO A NEW STATUS QUO

Bulgaria was an 'early adopter' of the communist movement, which pertained since the very start of the twentieth century and throughout WWII. Never, until 1988, had the country experienced a resistance movement to oppose Soviet-style state socialism (Baeva 2011). However, once the disintegration of the rest of the Soviet Bloc was raging in full force, the Central Committee of the Bulgarian Communist Party (BCP)—in a move orchestrated by the Soviets—voted for the resignation of the Chairman of the Presidium of the National Assembly, long-term leader Todor Zhivkov.

The Bulgarian Transition (also known as *prehodat* or *promenite*—'the changes') began on 10 November 1989. It brought about the rapid eradication of any communist presence in various spheres: villages, towns and cities carrying communist names were renamed, street names changed, new holiday systems adopted. Monuments from the communist era were replaced with those commemorating the victims of state socialism (Todorova 2009). Beyond the initial rage and implementation of symbolic gestures, the socioeconomic and political Transition brought fundamental changes to every domain of human life.

Lyudmila retired just before the onset of the Transition, aged 55. Together with her husband, they have been living in the same flat in central Sofia since 1957. However, with their sons and the grandchildren they helped raise now grown-up and many neighbours and friends deceased or bedridden, their focus has shifted to the countryside. They feel safer there, experiencing the city as an increasingly dangerous environment which no longer demonstrated the qualities they were expecting.

Stoyanka and her husband returned to the Village in 1994, when she retired. By that time, most factories in Gabrovo had been closed or sold off to foreign investors and too many workers had been laid off. Only eight years later, Stoyanka found herself struggling to adjust to widowhood, living alone and an environment increasingly difficult to navigate due to a diminishing pension, rising living costs and the decreasing availability of goods and services in the Village, her home. Her two daughters had built lives in Gabrovo. One granddaughter had gone to university abroad and got married there following graduation.

For today's older people in Bulgaria, the Transition often coincided with intimate and personal transitions in their lifecourse. Instead of 'only' having to adjust to a new stage of life—such as retirement or widowhood—many were confronted with the fundamental uprooting of their entire belief and value systems, having to work through the collapse of accustomed systems across all domains of life. The last decades of 'postsocialism'—a notion increasingly questioned by many for its (in)ability to describe socioeconomic systems and processes in former Eastern Bloc countries (Müller 2020)—were marked by continuing change and the establishment of a status quo of uncertainty.

Bulgaria went through a politically turbulent phase following the fall of the communist government in 1989. Possibly unsurprisingly, the country's first free parliamentary election led to the re-establishment of the Bulgarian Socialist Party (BSP) (the sequel to the Bulgarian Communist Party). What followed were two utterly unsettled decades, during which governments rarely lasted longer than two years (Baeva 2011). Throughout the 1990s, the power shifted between two successor parties every couple of years, and neither was successful in hiding that political leaders had spent much of their time stealing from, rather than working for, the people. The son of Bulgaria's last monarch—Simeon Sakskoburggotski (formerly known as King Simeon II)—reappeared to compete for the role of Prime Minister after a life in exile abroad. He won the 2001 election on the promise to end rampant corruption. His leadership lasted only four years, during which time he restituted crown lands to himself, among other acts of personal enrichment (Ghodsee 2008).

In 1991, Bulgaria officially began implementing the transition from a centrally planned to market economy and completed this process in 2002 (Vesselinov 2004). Just before the fall of state socialism, 12% of the national product could be attributed to agriculture and 61% to industry, demonstrating the success of efforts to industrialise the country during decades of socialist rule (Hoffman and Koleva 1993). Yet in shifting the focus from agriculture, Bulgaria also lost its competitive edge. Now, it had to demonstrate that it had capacity to cope with competitive pressure and that it had a functional market economy in place—requirements for joining the European Union. The country had agreed to a peaceful transition to market, beginning the restitution process (i.e., returning nationalised property to its original owners or their heirs), liberalising prices, abolishing controls on foreign trade, unifying the foreign exchange market and initiating labour relations towards tripartism (aka, institutional links between government, employers and trade unions) (Vesselinov 2004).

Restitution, that is, the return of property to its lawful owner in order to make up for the nationalisation or forceful dispossession during the communist era, started immediately following the fall of the communist government. Agricultural land was gradually returned to its rightful owners, either in its original place or in the same quantity and quality nearby. Furthermore, approximately 700,000 pre-1949 dwellings were returned by the end of 1999 (Miller 2003; Baeva 2011). Whilst many benefitted, large groups of people experienced restitution of urban and rural property as a massive disruption, leaving them homeless in its wake (Baeva 2011). People lost apartments that they had called home for decades, having to leave behind valuable memories and often close-knit networks. They had to find accommodation for themselves and their families in a newly emerging private housing market. This frequently meant being displaced from the centre to the edges of the city, from spacious turn-of-the-century apartments to new-built 'shoeboxes' on the outskirts.

> I am not one of [the poor pensioners] because they restituted my property, and my husband received his family's property back. So I let the properties and live from the income. I can afford to go to the concert, to the opera, to the theatre. (Violeta, 80, July 2019)

> I have lived here 25 years. We used to have a house here, but it was taken from us. My husband's parents had the house here, so when we got married in 1980 we joined them in the house. And then a few years later they came with the notice. Then we went to [the outskirts], into temporary housing, then when they completed building this compound we returned here. (Bonislava, 64, July 2016)

Not least because of the arrangements that needed to be made in order to accommodate displaced older people or younger families as a result of restitution, intergenerational co-habitation continues to be very common to this day. Frequently, moving in with members of the family coincided with one or both of the older generation spouses falling ill and requiring care, drawing on resources from the entire family:

> We used to have an apartment [in a street nearby], but they restituted it to its owners and we were allocated this one in exchange. At first, I let it because I lived with my son across the street, but then we wanted to live separately. (Albena, 86, July 2016)

> Once my husband fell ill my son moved in with me because I needed help taking care of him. It lasted a few years. (Milka, 78, July 2016)

Often, adult children and even grandchildren became unemployed and needed shelter. Economic restructuring had led to the liquidation of almost all industry, leaving a large portion of Bulgarians unemployed and lacking clear future outlooks (Baeva 2011). The country experienced a sharp decrease in economic growth and a substantial rise in unemployment (15% in 1993), leading to enormous strain on the social security system (Hoffman and Koleva 1993). For many of today's older people, the Transition came about when they were at the high point of their careers or around the time of their retirement. They were confronted with the sudden loss of a secure job at older age, sometimes only a few years from retirement. In a few exceptional cases, older people reported being unable to go into retirement formally, because the Transition had led to administrative chaos and the loss of their records:

> I am not a pensioner yet. I can't become a pensioner. They closed my institute, but nobody knows where our accounts disappeared. The bookkeeping disappeared. All my years ... I am 67 years old today. [...] Many years I have been saving, so I live on my savings. My husband has a good pension, so we live from his pension. But until they find the accounts, I won't be able to get my pension. We are living at the absolute minimum. (Boyana, 67, June 2016)

In other cases, the Transition tore gaping holes into the fabric of people's imaginations and expectations of a comfortable and safe retirement age. Depending on their professional and socioeconomic position prior to retirement, many recent pensioners deliberately took on 'post-retirement' jobs once they were formally retired. Some felt that in this way, they could continue to contribute to the emerging economy. For the majority, however, working post-retirement was a nothing short of a necessity in order to make ends meet:

I was afraid to become unemployed during the Transition, so I [left the state-owned enterprise I worked with] and started work with an insurance company. But this was swallowed and so effectively I became unemployed. After a month I found work with a foreign firm. I've been with them 19 years now. I could have retired nine years ago, but I stayed on. I like my work, we're a good team. It's enjoyable. The moment you go to work and don't want to, you should stop working. (Bonislava, 64, July 2016)

I retired in 1991—they used to retire us much earlier back in the days. I stopped working after I retired, but then a few years later I started working again with my previous patients. (Violeta, 80, July 2016)

I have been retired for 15 years [since 2000]. I stay at home now, I don't work. But I used to work after retirement, for 10 years, between what, 55 and 65. I am an accountant, but I used to work as a cleaner. (Tsvetelina, 72, July 2016)

For those on the brink of retirement or recently retired, the Transition amplified the inequalities resulting from the collapse of state socialism and its associated systems of welfare. Depending on the type of past employment, being older and more experienced presented an advantage for some in well-respected professions or already in positions of leadership; others, however, found themselves unable to compete in an emerging job market rooted in the principles of neoliberalism. They had to accept jobs that did not, in any way, match their education, experience or skill sets. In this way, the Transition exposed and intensified already existing inequalities based on gender, class and education, among others.

Older women were increasingly charged with the care of grandchildren, frequently at the expense of early retirement or the loss of their post-retirement jobs, sometimes leading to the deterioration of their own circumstances. Today, women generally receive lower pensions than men. Although Bulgaria was one of only five European countries with gender parity between, for instance, men and women scientists (Kağitçibaşi 2008), in 2019 women aged over 65 received pensions that were on average 22.8% lower than the—already small—pensions of men (eurostat 2020a). Take Raina, aged 68, who worked for the statistical office for almost three decades. When state socialism fell, she became unemployed and opened a little shop in order to earn an income. However, she became a grandmother in 2003 and had to stay home to look after her grandchild so her daughter could go back to work. These days, Raina helps looking after her small granddaughter, travelling to her daughter's home on the outskirts of the city. She works as a cashier at the local supermarket:

> We retired [in the early 2000s] at a time when our pensions were still sufficient. Then things changed quickly. Calculations were made in a very strange way. None of our premiums were taken into account ... Anyways, pensions are low now. Very low. (Raina, 68, September 2016)

> The pensions are small. But I know how people live in Europe, and how our pensioners live. 46% below the poverty line! [...] People have to count their change! I have a friend, a great doctor, who had to go through the bins to look for food after retirement. (Violeta, 80, July 2016)

The lifting of the 'triple burden' (Einhorn 2002) of work, family and community in women's lives during the Transition to a market economy and liberal democracy was interpreted by some women as liberation, as the freedom to stay at home and be a housewife (Todorova 1993; Ghodsee 2004). However, the Transition did affect women disproportionally through factors such as shrinking pensions and evaporating social services. 'Freedom' came hand in hand with the decrease of state investment in public services—including childcare—that was seen as necessary in order to meet International Monetary Fund constraints and be able to join the European Union (Ghodsee 2004). It carried implications for Bulgaria's women, who now had to divide scarce time between work and caring for children, the elderly and the sick where they could not afford to stay home.

In many cases, caring for the elderly (their own parents) required the return to the home town or village. Konstantinov and Simić (2001) note that post-1989 Bulgaria's rural-urban reciprocity 'can be observed in the return to the villages of "young pensioners" who have been replacing their very old or deceased parents, leaving their urban apartments to their children'. However, whilst pensioners returned to the countryside, hundreds of thousands of young Bulgarians migrated from the countryside to towns and cities, and later to richer countries in Europe and elsewhere (Baeva 2011). Unemployment rates in 2016 were 6.3% and 12.4% for urban and rural Bulgaria, respectively (NSI 2017). Bulgaria had 430,000 movers of working age (24–65), of which 103,000 were resident in Germany, 98,000 in Spain and 73,000 in the United Kingdom (Fries-Tersch et al. 2017). The local workforce stood at approximately 3.1 million (NSI 2020). This means that an estimated minimum of 11.5% of Bulgaria's workforce were abroad, within the limits of the European Union, with migration further afar not even taken into account. Of course, such high levels of emigration are reflected in the composition of the household: in 2005, 22.5% of older people above the age of 65 lived alone; this proportion had climbed to 40.3% in less than 15 years. Women are particularly affected by this transition, with almost one in two women above the age of 65 living alone in 2019 (eurostat 2020b).

The rampant emigration of people of working age has consequences with regard to the decline of family-based elderly care; in addition, leaving children behind to be looked after by their grandparents poses extra burden on already strained resources (Kulcsár and Brădățan 2014). Among older people ageing alone in rural settings in Bulgaria, Conkova et al. (2019) find that social hardships include lack of intergenerational care—with regard to both giving and receiving—and feelings of abandonment. Feelings of loneliness 'are engendered by the departure of […] children and grandchildren, either abroad or to distant cities within the country. On the one hand, this corresponds with the theoretical notion of loneliness being determined by the composition, size and width of individuals' social networks; on the other it hints at the role of normative expectations in Central and Eastern Europe' (Conkova et al. 2019).

## EXPECTATIONS, IDENTITIES, ROLES: FINDING AGENCY UNDER UNCERTAINTY

In this section, I work through the interconnected and overlapping domains of expectations, identities and roles of older people in Bulgaria and how they are situated in the context of the country's persistent rural-urban reciprocity (Smollett 1985; Konstantinov and Simić 2001). In the city, as in the village, older people's expectations are that the younger generations should look after them financially, physically and in every other way. Although they are trying to delay becoming 'a burden' for as long as they can, they expect that—should worse come to worst—children and grandchildren would be there to look after them. These expectations are rooted in long-established norms as part of larger value, belief and norm systems; they underpinned their own actions in the past, when they returned from the city to the village or moved in with their parents or in-laws to provide care for them. However, in many cases, these expectations are not met in the face of shifting social norms. Take Albena: she lives alone, her husband died in 1989. She has two children in their 50s and 60s, as well as four grandchildren and a great-granddaughter. Albena is not in good health. She has difficulties walking and therefore tries to reduce her ventures outside the flat to a minimum. She uses the dining room as a bedroom in order to save on heating costs. She does not like taking money from her relatives, but relies on them for financial help when needed.

> I do think that the younger generations should look after the elderly. But unfortunately, this doesn't happen very much. We don't even get the respect we deserve. From my children, I don't get it to the degree I would have hoped. I have tried all my life to make sure they are well educated, that they study, study, study. Characteristically for Bulgarians, I have tried to make life easier for them, to take things off their plate [...] They have gotten used to receiving, only [...] I admit this is partly my fault, I have not educated them as I should have. (Albena, 86, July 2016)

Like her, many feel disappointed by the mismatch between their expectations and the reality of ageing and how the younger generations treat them. They feel hurt by the lack of esteem and respect they receive, mentioning how this manifests in the absence of the smallest of gestures, such as being offered a seat on public transport. Many take a stance and categorically declare that they do not expect anything from the younger generations and that they are perfectly capable of taking care of themselves without help, financially and otherwise. Then, some are quick to defend their adult children, justifying the situation with the adverse economic conditions and the assumption that the younger generation has to look after themselves, first.

> I don't blame them. It's not their fault. They are trying to survive. (Zora, 70, July 2016)

With regard to notions of identity, political affiliation and class seem to play much more decisive roles in the formation of the contemporary identities of older people than, say, gender, and just where on the spectrum older people position themselves seems to vary with their geographical location and rural or urban identity. Despite being an agrarian society until not too long ago, Bulgarian popular culture has always had a dubious relationship with the countryside (Staddon 2004). The dichotomy of the communist-era ideal of the urban industrial worker and the crafty rural peasant farmer did little to undo the widespread superciliousness of urbanites looking down upon rural identities and lifestyles. Attitudes persisted tenaciously notwithstanding the critical importance that the agricultural sector took on in the 1990s, when other sectors were hard hit by the Transition and the restructuring that came with it. More recent identity associations assign anti-communist (and thus 'progressive') views to the city—and communist political convictions to the countryside (Staddon 2004). My findings indicate that, at least among the people that took part in this study, such associations do hold true, and I discuss possible reasons why in Chap. 3.

Indeed, in the Village, only few older people do not identify with the left or far-left. They have been members of 'the party' (the Bulgarian Communist Party and later Bulgarian Socialist Party) for as long as they can remember. Many are disappointed that their children and grandchildren have drifted politically to the right. They are particularly upset about the newly introduced educational framework which portrays the socialist era as the dark ages of Bulgarian history.

> We managed to buy cars, to build homes … Now, they work and work and work and they can't manage to achieve what we have. (Nedelya, 69, July 2016)

In the city, too, many older people today appreciate the advantages of the 'old system' when looking back:

> There were good things in the old days. Everyone had a job, things were getting built, and things were progressing. Of course, not everybody liked how things were going, so they revolt. Great—but why […] not keep what was good and […] make things better? Back in the days, there was not a place where people didn't grow things. And now? They closed all factories, all agriculture. (Boyana, 67, July 2016)

However—apart from a short period of enthusiastic royalism, when the former King Simeon II took power as Bulgaria's Prime Minister at the beginning of the millennium—the majority of older people in the city appear to have grown indifferent to politics, and although they no longer wish for the communist days to return, they are disappointed with the changes that have taken place. Blagun (66), for instance, continuously compares current conditions in the country with those under state socialism. He was born in a village in 1950 and moved to Sofia when he was eight years old. Like his father, he worked for 'the' train factory (there only was one). Entirely in contrast to his political convictions and on the advice of his mother, he became a member of the party over practical considerations, specifically expectations of progress in his job and advantages when it came to the allocation of housing or goods. Following a complicated residential history, Blagun now lives in a 1980s flat in central Sofia, having had to surrender his villa in the countryside to his adult daughter and her family—who had nowhere else to go. He retired early in 2005, after a health episode that left him unable to work, and earns some extra income doing occasional shifts at a coffee shop nearby. Blagun feels increasingly alienated by people's materialist views. He is convinced that, were he young now, he would do what the younger generations are doing: pack his suitcase and leave.

> There is no point staying in this country. […] I don't want to go back to the old days, it was ridiculous back then. But it's the opposite now. From one extreme to the other. (Blagun, 66, July 2016)

The underlying disappointment with their current position is mirrored in older people's responses to the request to self-identify with a social class: most are quick to state that social class no longer means anything at all; that Bulgaria has changed and that now, there are only two classes, the rich and the poor. And that clearly, older people, regardless of their level of education, belong to the latter. Older people are 'third class' citizens. 'That's capitalism for you', they often say in the village.

> We are the lowest class of people. The people who don't have any ambitions at all. There is nothing I can do. Nothing. I can have a coffee, smoke a cigarette or two, and that's it. I want to go to the cinema—I can't. I want to go to the opera—I can't […] It's 22 BGN. I can't afford that. (Boyana, 67, June 2016)

Behind the potently expressed sense of powerlessness woven into the narratives of older people like Boyana is the combination of multiple factors. They include the general feeling of having lost (or being cheated out of) a socio-political system that, albeit imperfect, was still superior to that replacing it; the severe restrictions older people face due to the often trifling financial means they have access to; or simply 'age' as an increasingly dominant aspect of their identity, tied to declining health and diminishing ability to exert an influence on decisions across different scales, from the familial to the national. For instance, most older people in the city assume that their opinion is irrelevant and that they have no way to effect political decision making. They do not think that they as a group should even be considered in the design of future policies. Whilst they do feel the obligation to vote, most admit to feeling distrust towards the outcome of voting, and towards the government more generally.

> In society, I don't think we matter. I feel we are only used when it's time to get our votes. Else, they forget we exist. I don't think if there are institutions or organisations that I would turn to if I should be or get in trouble. I don't think our jurisdictive system is functioning well. I have children, I rely on them. I turn to them when I need help. (Lyudmila, 86, July 2016)

In contrast, many report that within their family they feel respected and that their opinion is sought out, but they do not expect adult children to seek their advice. Although they feel that they should stay out of younger generations' lives as much as possible, they feel that they have much to contribute based on their experience. However, shifts are taking place in older people's relationships with younger people of working age as the nation's economy remains stagnant, at best:

> Back in the days, we used to look after the parents. We got 100-120 BGN salary, and regardless, we didn't lack anything. We had everything we needed. No luxury, of course. But we had everything we needed and we looked after the parents. That's no longer the case. They're unemployed. Instead of looking after us, they come to us and ask us to help them. That's not normal. (Boyana, 67, June 2016)

There is the feeling of responsibility among those who have regular contact with grandchildren, the feeling that they have to teach them how to behave well. Simultaneously, they acknowledge that relationships with adult children and grandchildren are good precisely because generations are no longer habitually co-habiting, and that relationships could be strained should the opposite be the case. In the village, most are happy to see their adult children and grandchildren making their own decisions, but some feel disrespected and even threatened by high levels of independence. They want to be consulted. A few take precautions to make sure they have a say. They sign legal contracts with their relatives which outline their rights; for instance, their right to live in and be cared for in exchange for transferring property rights to their relatives.

When looking back at their lives, older people rarely describe situations of decision making or choice. In their narratives, they use language that almost exclusively indicates the agency of others, primarily the state. For instance, they were 'sent to the city to study', they were 'given a job in a state-owned company', they were 'sent to work in the countryside', they were 'given an apartment' and they were 'allocated a car'. Decisions were made for them and required them merely to do as they were told. Many had internalised such an attitude to a degree that led them to transfer decision-making responsibilities to their adult children or even grandchildren following the evaporation of the state from public and private lives with Transition:

> I think my opinion shouldn't be taken so seriously. I think my son should be making the decisions. His decision is going to be more appropriate than mine. (Miroslav, 81, June 2016)

> If [the children] don't need me to look after [my granddaughter], if they have given me permission, I immediately go to the villa. (Gergana, 74, July 2016)

Only few use a language that indicates some initiative or involvement in decision making and following through, and then agency usually takes the form of peeling away layers of state (or party) expectations:

> I used to be a Komsomol secretary. [Then] I started work and decided that for me it's better to focus on my family and my children's development, instead of staying with the party. (Bonislava, 64, July 2016)

Often when individual agency was reported, older people referred to the most recent Transition decades of their lives. Furthermore, the decisions in question were relatively minor, such as whether or not to participate in organised activities at the pensioners' club or what to cook:

> So for instance I decided [my daughter-in-law] should fill peppers with feta cheese and eggs, so I went and bought the peppers. I will grill them, I will

peel them, but she knows how to fill them. She's a very good cook. (Milka, 78, July 2016)

Overall, however, others were and sometimes still are making decisions for older people today, or on their behalf. On the one hand, it can be argued that this apparent lack of agency is fully in line with what would be expected based on the political upbringing they received as well as the totalitarian structure within which they lived the majority of their working lives. The state did, indeed, take care of everybody and everything. On the other hand, as previously stated, I report here mostly on the lifecourse narratives and experiences of older women; therefore, it could be argued that gender—and the oppression of women, in particular—has a role to play in whether or not people draw on and display agency. However, as Ghodsee (2004) rightly points out: 'Certainly, Bulgarian women's attitude toward their own abilities and their place in society is still very much defined by socialist ideas about the equality of the sexes'. Older women did not consider themselves victims of a male-dominated society, nor did they think of their agency as oppressed on the basis of their gender.

The lifecourse perspective poses that individuals have agency to construct their lives. The concept of agency denotes 'the capacity that an individual acquires to plan his/her own future assuming an active, conscious and intentional role in achieving this future' (Romaioli and Contarello 2019). The boundaries within which the individual as an active agent can make decisions are provided by 'family background, stage in the lifecourse, structural arrangements, and historical conditions' (Bengtson et al. 2005). Agency in ageing is thought as 'the methods that [older people] adopt to delay growing old and to address the decline linked to age' (Romaioli and Contarello 2019). As we have seen in Chap. 1, this conceptualisation is driving policy agendas, including the WHO's 'age-friendly cities' and 'active ageing' initiatives. However, as Wray (2003) notes, the uncritical application of such theoretical frameworks 'across ethnic and cultural diversity' risks discarding differences in the experience of ageing across ethnic (and cultural) boundaries.

Precisely because of the possible differences in value, belief and norm systems across different cultures and ethnic groups, applying a universal framework may render ageing and older people in settings that are not Western European or North American inert and vulnerable—where, in fact, within their life-worlds and their own perception they may be highly adaptive and resilient, responding to continuously changing circumstances to construct and achieve desired futures. It is therefore urgent to recognise the unique character of ageing in the Global East (Müller 2020).

In the case of Bulgaria, a member of the European Union on the very margins of Europe, this means ageing in the context of uncertainty emerging from a complex palimpsest of political melee, contrasting cultural influences and ever-transforming value, norm and belief systems. The state and its former promise to take care of everybody and everything seem long forgotten, and older people hardly ever mention it in relation to their present. If they do, they do not try to hide their disappointment:

> I think the state should have made sure that our pensions are big enough so that the young can live their lives and so that we are financially independent. (Svetla, 76, July 2016)

Under the country's five decades of socialism, the decisions that individuals could make took place within comparatively narrow and tightly delimited spaces. The kinds of dependency and social contact imposed by the socialist state stand in stark contrast to autonomy and independence as indicators of agency within a liberal Western value system. Older people today have had to make the painful transition between these two ends of the continuum over a short period, with profound implications for their expectations, identities and roles embedded within a larger and perpetually shifting social order. Only few have been successful in finding agency under uncertainty.

## Conclusions

In this chapter, I have attempted to weave a narrative about the ways in which social and historical context are entangled with the lifecourse of older people in Bulgaria, leading up to the present day and the ways in which older people position themselves within an ever-changing social order. Methodologically, doing so has required the analysis of older people's responses to questions derived from the human ecosystem framework around longer-term social cycles (the lifecourse) and social order (characterised by the elements of identity; norms and hierarchy). Whilst the human ecosystem provided the frame for data collection, data analysis has been guided by the lifecourse approach and its principles. The emerging narrative is structured historically by the eras of state socialism (1944–1989) and the Transition (since 1990) and the conceptual domains of migration, housing and labour. The lens of agency allows to unpack ongoing transitions in the country's social order in locating older people's normative expectations, shifting identities and changing positionalities as part of wider social hierarchies.

Older people today have lived through a diverse set of historical periods with profound impact on their lifecourse: they were born around World War II, grew up and entered the workplace under state socialism, experienced the turbulent Transition period and are now faced with a distinct form of savage capitalism (Ghodsee 2005, 2011; Müller 2020). As Hall (2019) argues, such transitions can be 'personally affective, they can lead to rupture, fragmentation and disjuncture, having lasting personal and relational impacts'.

Under state socialism, the state took care of everybody and everything, consolidating a kind of socialist paternalism (Verdery 1996) that, in its essence, severely restricted individual agency and choice. Bulgaria's five decades of state socialism marked the nation's era of urbanisation, which found expression in significant, yet initially tightly controlled, rural-to-urban migration. Socialist housing complexes in Bulgaria's cities mushroomed as a response to growing urban populations and the already existing rural-urban reciprocity was consolidated through measures such as the second home, or villa. Gender relations were fundamentally transformed, freeing up women so that they could fully participate in the economy and, importantly, contribute to the progress of socialism. The norm of the multi-generational household meant mutual support between generations, with the elderly helping with childcare and younger generations looking after the elderly in return.

The Transition period brought about an era of uncertainty that required individuals to adjust a great many aspect of their lives. Many older people who were working in cities chose to return home to the countryside— often to care for elderly parents. Simultaneously, the period was marked by the continued rural-to-urban migration of young people and the unprecedented outmigration of young people abroad. In the city, the restitution of pre-1949 properties to their rightful owners saw those occupying restituted properties lose their homes and having to negotiate accommodation in a newly emerging residential market. For many, this meant moving (back) in with adult children or grandchildren in order to save costs. The Transition often meant unemployment or early retirement for many in their prime working years in the early 1990s. They were thrown into a newly liberalised job market, often lacking the necessary skills to thrive in an increasingly competitive environment. For those on the brink of retirement or just retired, the collapse of socialist welfare systems meant that they had to make do on pensions that—to this day—can only be described as an insult to anyone with a history of employment. Post-retirement jobs became the norm, exposing deep-rooted inequalities along the lines of gender, class and education and affecting predominantly women.

Bulgaria's elderly have had to adjust their value, norm and belief systems to the ever-changing realities of collapse and rebuilding and, eventually, a permanent mode of uncertainty. Their social-norm-driven expectations of the younger generation (to take care of them as they age, financially and otherwise) cannot be met under conditions of a struggling local economy, rampant outmigration of the younger generations and shifting values in an increasingly materialist society. For older people, identity markers such as political orientation and class have taken on importance. Powerlessness and marginalisation are woven into older people's narratives across these domains, underlining their seeming willingness to hand responsibility to others.

I argue that Bulgaria's systemic socio-political transitions over the last 80 or so years have shown enormous impact on people's lifecourses, including the ways in which older people position themselves and act in society today (see Chap. 3). The experiences of life under socialism in the People's Republic of Bulgaria have, on the one hand, taught older people to accept that they cannot influence the course of their lives in major ways, assigning this responsibility to the state. On the other hand, the Transition, following the fall of state socialism, was a hard hit on most people of working age at the time, taking away the state-provided safety net and exposing them to new disruptions such as unemployment, homelessness and the urgency of learning how to navigate the ever-changing rules of an emerging state. Eventually, over the last two decades people have learned to expect nothing from the state; to rely on their close social networks, including family and friends; and, where this is impossible because of migration experiences or intergenerational conflicts, to rely on themselves. As I will discuss in detail in Chap. 3, their everyday lives today are shaped by uncertainty, compelling them to draw on a newly developed agency to navigate the status quo.

# REFERENCES

Baeva, I. (2011). Dve desetiletiya bŭlgarski prekhod—predpostavki, problemi, ravnosmetka [Две десетилетия български преход—предпоставки, проблеми, равносметка]. *Drynovs'kyy zbirnyk [Дриновський збірник]*, 4(2011), 330–341.

Barron, A. (2019). More-than-representational approaches to the life-course. *Social & Cultural Geography*, 1–24. https://doi.org/10.1080/1464936 5.2019.1610486.

Bengtson, V. L., Elder, G. H., & Putney, N. M. (2005). The Lifecourse perspective on ageing: Linked lives, timing, and history. In M. L. Johnson (Ed.), *The Cambridge handbook of age and ageing* (Cambridge handbooks in psychology) (pp. 493–501). Cambridge: Cambridge University Press.

Chan, K. W., & Zhang, L. (1999). The *Hukou* system and rural-urban migration in China: Process and changes. *The China Quarterly, 160*, 818–855.

Conkova, N., Vullnetari, J., King, R., & Fokkema, T. (2019). "Left like stones in the middle of the road": Narratives of aging alone and coping strategies in rural Albania and Bulgaria. *The Journals of Gerontology: Series B, 74*(8), 1492–1500.

Einhorn, B. (2002). *Cinderella goes to market*. London: Verso.

eurostat. (2020a). *Closing the gender pension gap?* Brussels: eurostat.

eurostat. (2020b). *Distribution of population aged 65 and over by type of household—EU-SILC survey [ilc_lvps30]*. Brussels: eurostat.

Federal Research Division. (2017). *Bulgaria country studies*. Retrieved February 22, 2018, from http://www.country-studies.com/bulgaria/.

Fei, X. (1992). *From the soil, the foundations of Chinese society: A translation of Fei Xiaotong's Xiangtu Zhongguo, with an introduction and epilogue* (G. G. Hamilton, & W. Zheng, Trans.). Berkeley, Los Angeles, and London: University of California Press.

Fries-Tersch, E., Tugran, T., & Bradley, H. (2017). *2016 annual report on intra-EU labour mobility*. Brussels: European Commission, Directorate-General for Employment, Social Affairs and Inclusion.

Gaubatz, P. (1998). Understanding Chinese urban form: Contexts for interpreting continuity and change. *Built Environment, 24*(4), 251–269.

Ghodsee, K. (2004). Red nostalgia? Communism, women's emancipation, and economic transformation in Bulgaria. *L'Homme: Zeitschrift für Feministische Geschichtswissenschaft, 15*(1), 23–36.

Ghodsee, K. (2005). *The red Riviera: Gender, tourism, and postsocialism on the Black Sea*. Durham and London: Duke University Press.

Ghodsee, K. (2008). Left wing, right wing, everything: Xenophobia, neo-totalitarianism, and populist politics in Bulgaria. *Problems of Post-Communism, 55*(3), 26–39.

Ghodsee, K. (2011). *Lost in transition: Ethnographies of everyday life after communism*. Durham and London: Duke University Press.

Hall, S. M. (2019). *Everyday life in austerity: Family, friends and intimate relations*. Cham: Palgrave Macmillan.

Hoffman, M. L., & Koleva, M. T. (1993). Housing policy reform in Bulgaria. *Cities, 10*(3), 208–223. https://doi.org/10.1016/0264-2751(93)90031-D.

Hopkins, P., & Pain, R. (2007). Geographies of age: Thinking relationally. *Area, 39*(3), 287–294.

Ivanov, R. (2007). Political economy during the epoch of socialism (1949-1989) [Narodno stopanstvo prez epochata na socialisma (1949-1989 g.)]. *Dialog, 2007*(4), 116–150.

Kağitçibaşi, Ç. (2008). Caution: Men at work. *Nature, 456*, 12. https://doi.org/10.1038/twas08.12a.

Konstantinov, Y., & Simić, A. (2001). Bulgaria: The quest for security. *Anthropology of East Europe Review, 19*(2), 21–34.

Kulcsár, L. J., & Brădățan, C. (2014). The greying periphery—Ageing and community development in rural Romania and Bulgaria. *Europe-Asia Studies, 66*(5), 794–810. https://doi.org/10.1080/09668136.2014.886861.

Machlis, G. E., Force, J. E., & Burch, W. R. (1997). The human ecosystem part I: The human ecosystem as an organizing concept in ecosystem management. *Society & Natural Resources: An International Journal, 10*(4), 347–367.

Miller, L. R. (2003). Land restitution in post-communist Bulgaria. *Post-Communist Economies, 15*(1), 75–89. https://doi.org/10.1080/1463137032000058395.

Müller, M. (2020). In search of the global east: Thinking between north and south. *Geopolitics, 25*(3), 734–755. https://doi.org/10.1080/14650045.2018.1477757.

NSI. (2017). Labour_3.2.3_EN.xls. In L. M. S. Department (Ed.). Sofia.

NSI. (2020). *INFOSTAT*. Retrieved June 20, 2020, from https://infostat.nsi.bg/infostat/.

Parusheva, D., & Marcheva, I. (2010). Housing in socialist Bulgaria: Appropriating tradition. *Home Cultures, 7*(2), 197–215. https://doi.org/10.2752/175174210X12663437526214.

Romaioli, D., & Contarello, A. (2019). Redefining agency in late life: The concept of 'disponibility'. *Ageing and Society, 39*(1), 194–216. https://doi.org/10.1017/S0144686X17000897.

Smollett, E. W. (1985). Settlement systems in Bulgaria: Socialist planning for the integration of rural and urban life. In A. Southall, P. J. M. Nas, & G. Ansari (Eds.), *City and society: Studies in urban ethnicity, life-style and class*. Leiden: Institute of Cultural and Social Studies, University of Leiden.

Smollett, E. W. (1989). Life cycle and career cycle in socialist Bulgaria. *CULTURE (Journal of the Canadian Anthropology Society), IX*(2), 61–76.

Staddon, C. (2004). The struggle for Djerman-Skakavitsa: Bulgaria's first post-1989 "water war". In *Drought in Bulgaria: A contemporary analog of climate change* (pp. 289–306). Aldershot: Ashgate Press.

Staddon, C., & Mollov, B. (2000). City profile: Sofia, Bulgaria. *Cities, 17*(5), 379–387. https://doi.org/10.1016/S0264-2751(00)00037-8.

Todorova, M. (1993). The Bulgarian case: Women's issues or feminist issues? In N. Funk & M. Mueller (Eds.), *Gender politics and post-communism: Reflections from Eastern Europe and the former Soviet Union* (pp. 30–38). New York and London: Taylor and Francis.

Todorova, M. (2009, 9 November 2009). Daring to remember Bulgaria, pre-1989. The Guardian. Retrieved from https://www.theguardian.com/commentis-free/2009/nov/09/1989-communism-bulgaria.

Verdery, K. (1996). *What was socialism, and what comes next?* Princeton, NJ: Princeton University Press.

Vesselinov, E. (2004). The continuing 'Wind of Change' in the Balkans: Sources of housing inequality in Bulgaria. *Urban Studies, 41*(13), 2601–2619. https://doi.org/10.1080/0042098042000294583.

Windle, G. (2011). What is resilience? A review and concept analysis. *Reviews in Clinical Gerontology, 21*(2), 152–169.

Wray, S. (2003). Connecting ethnicity, agency and ageing. *Sociological Research Online, 8*(4), 165–175. https://doi.org/10.5153/sro.866.

Yoveva, A., Gocheva, B., Voykova, G., Borrisov, B., & Spassov, A. (2000). Sofia: Urban agriculture in an economy in transition. In *Cornell food and nutrition policy program working papers* (Vol. 54, pp. 501–518). Ithaca: Cornell Food and Nutrition Policy Program.

CHAPTER 3

# Doing Everyday Life: Patterns, Resources and Adaptive Mechanisms

**Abstract** Iossifova draws on observation and interviews to demonstrate how Bulgaria's abandoned older people deploy newly found agency to cope with uncertainty in their everyday. She explores their patterns of everyday life in the city and in the Village before turning to discuss the domains of health and energy and present older people's accounts of managing meagre pensions, sustaining ailing bodies and making do in tattered homes. Iossifova argues that older people in the city and in the Village rely on vastly different resources and infrastructures and that they display astonishing resilience in the face of adverse financial, material, physical, political, social and cultural circumstances, regardless of their geographic location.

**Keywords** Bulgaria • Post-socialism • Older people • Everyday life • Housing • Health • Energy • Pensions • Infrastructure • Resilience • Rural-urban reciprocity • Adaptive mechanisms • Uncertainty

They say there was this bad smell in the stairwell. It crept up on them from nowhere and kept getting stronger. For weeks they wondered where it came from. This smell of decay. As if something had been left out by mistake, some food stuff someone forgot to put away. That smell of rotting. At first, they thought it was a dead rat, or a mouse, or a cat. They had that once, a squirrel had found its way into the cellar, got trapped and made that terrible noise, keeping the building awake all night. But this was different. They kept talking about the smell in passing, when bumping into each other, gasping for air, up the stairs, down the stairs. The lift had been out of service for a while. And then the children noticed. It came from the fifth floor, they said. The janitor had to break into the apartment, and there they found her. On the sofa. No one had seen her in months. Since the lift had stopped working, actually. The food cupboard was empty. The phone was dead. She just sat there, as if waiting for the heating to finally come on. It had been disconnected; for too long, the bills had not been paid.

*    *    *

It is mid-morning on a winter weekday. The place is never packed around this time. Italian pop fills the room, competing with the happy chatter of four well-dressed ladies in their 70s. They sit around their usual table, their coats insouciantly thrown over the back of their chairs. Fur trims showing here and there; the gold and silver straps of handbags sparkling in the light of carefully positioned lamps. The flat screen TV above them playing endless loops of documentaries on Italian cuisine. The smell of melted cheese making its way from the kitchen contesting the subtle fragrance of their perfumes. They talk in-between sips of lattes, cappuccinos and freshly squeezed orange juices. Imported prosciutto, tomatoes on the vine. Their breakfast sandwiches prepared to their liking. The staff greeting them by their name. On days like these, what with the heavy snow and all, their taxis drop them at the entrance. Their hair is made up nicely, the skin on their faces recently lifted and tight, their nails varnished. The smartphones on the table buzzing with fresh news from their sons, daughters and grandchildren, a short plane ride away in Germany, France or the States.

*    *    *

The air is crisp, the wind is cold. The barren birch trees bend ever so slightly, interrupted by the barking of hungry stray dogs. The sky is grey, the crumbling facades of the three-storey buildings blend almost seamlessly with their surroundings. A small woman, wrapped in an oversized jumper and heavy scarf, stands by the enormous silver garbage bin. Her boots are too big. Too big and too, too old. There is a layer of slippery ice underneath the thick coat of snow. The streets, the parked cars, the side-walks, the green areas of the past, they all melt into one, white and grey and brown. In one bare hand, she carries a plain plastic bag, perforated with age. With the other, she holds on to the handle of the bin's lid. She looks around, as if to make sure no one is watching, then uses both hands to remove the weighty lid. She scans the contents of the bin, then closes her eyes in disappointment. With a hint of despair, she glances down the long, tree-lined street as she removes a strand of silver hair from her face. Nothing. She is too late. Everything is white. White and grey and brown. It will be a long walk home. And she will still be hungry. There is a layer of slippery ice underneath the thick coat of snow.

* * *

The tram. Again. Sitting behind the counter, she opens the drawer. Again. One, two, three, four ... This will hardly be enough to pay for her pre-scription. The bottle is empty, she needs to buy her medicine today. The pharmacy is still open. Two more clients, maybe three, that would be enough. She looks down her knitted vest, down her swollen legs; the veins popping out, the slippers eating into her flesh. She closes the drawer. She's read the newspaper twice now, from beginning to end. Her neighbour left it when he stopped by in the morning. He knows she likes to read the news. Now that her son took the TV. She stares at the empty plastic cup, shrunken from the hot black coffee she had in it in the afternoon; the sugar still sticking to the wrinkled bottom. The tram passes, breaks squeak-ing. Maybe now? One, two, three, four ... One-by-one they get off. She nods as she makes an effort to smile at the middle-aged red head. It's that time of the day. Again. It seems people have stopped buying from small shops like hers. It seems it's all about the big chains, the super markets, the malls. The pharmacy will close soon. She opens the drawer. One, two, three, four ... It will not be enough.

* * *

Tick, tock. Tick, tock. The window is ajar, swinging back and forth as the wind blows ever more cold air into the small chilly room. He lies on his side, drawing the thin wool blanket over his mouth and nose with one trembling, wrinkled old hand. His knobby knees are pulled up to his chest, his glance indifferent as he surveys the heap of snow that has collected underneath the window. The wooden planks seem to bend under the weight of his bed. There are bright marks in the corner where once the oven stood. A round black hole above the sink where once the stovepipe disappeared into the wall. Cold water dripping from the tap. Some hard old bread and cheese forgotten on top of the newspaper on the floor. Tick, tock. Tick, tock. The sun will soon be gone. Get up, or not? Tick, tock. No one will come today. Tick, tock.

\*    \*    \*

There is the sound of children laughing; the clinking and tinkling of cutlery as the younger women set the table. The men are out of sight behind two or three rows of tomato plants, but the smell of charcoal and freshly barbecued meat cannot be ignored. It is this time of the day, just before sunset, the windows are lit from inside the small house. No one has switched on the light on the terrace yet. The garden is huge. She's done well this year—peppers, beans, tomatoes. Some really nice cucumber, too. It is a bit cramped when they are all here, but fun. Upstairs the great-grandkids, the granddaughter with her husband; downstairs, the grandson and his family—all the way from Belgium. Her daughter comes home from the city, too. For six weeks they come, all of them, and it's like in the old days, when she was a child. In spring, they paid to have the roof done and the facade painted. She sits at the big rectangular table and watches the hustle-and-bustle. It's good to have them here.

\*    \*    \*

The paragraphs above are based on fleeting observations, short visits and brief conversations in the course of my fieldwork. I include descriptions of these moments here because they illustrate the very personal and highly relational experience of ageing in a context characterised by perpetual change, deep-seated inequalities and a state of permanent uncertainty. As discussed in Chap. 2, Bulgaria's past and ongoing systemic socio-political transitions have greatly influenced people's lifecourses. To a large extent, today's older people have normalised uncertainty as the new status quo. They have activated a range of adaptive mechanisms in response to anticipated and ongoing disturbances of economic, social, cultural, political and highly personal character—as well as disturbances related to extreme events like heat waves, cold spells or floods. Older people implement adaptive mechanisms as strategies and tactics in the smallest of actions in their daily lives. Patterns of everyday life are therefore shaped by complex relationships between older people and their environment, including the social, the physical and the virtual environments, as I discuss in Chap. 4.

Here, I examine the ways in which older people position themselves to enact adaptive mechanisms in the context of the permanent state of uncertainty gripping Bulgaria today. I work through the narratives of older people in Sofia and in the Village to relay detailed accounts of their everyday lives and how these are entangled with different aspects of the human ecosystem framework (Machlis et al. 1997, see Chap. 1). Although I touch upon a series of human ecosystem categories and elements, those explicitly underpinning the material presented in the following sections include *social cycles* (at the scale of the everyday as well as the four seasons), *social institutions* (health), *socioeconomic resources* (capital, information, population), *natural resources* (land, energy) and *cultural resources* (organisation, beliefs). Theories of change and adaptation support my analysis and interpretation of above interactions.

In the following sections, I first present common theories of change, adaptation and coping developed and applied in the study of ageing. I then illustrate the patterns of everyday life in the city (Sofia) and in the Village and work through some key elements of everyday life, including financial resources, health and living conditions, unravelling once more how shifts in political and environmental circumstances can impact on the lives of individuals. I unpeel some of this complex layering to extract a common thread: as they age, people in Bulgaria have come to rely on a tightly knit community of family and friends to adapt to challenges under uncertainty. I conclude that any intervention designed to improve their ageing experience should therefore consider and support re-emerging intergenerational systems of care. I argue that in the face of dramatic and all-encompassing transformation affecting every domain of their lives, Bulgaria's elderly appear to be surprisingly resilient.

## THEORIES OF CHANGE AND ADAPTATION IN AGEING

The resilience of human systems—social resilience—is defined as 'the ability of groups or communities to cope with external stresses and disturbances as a result of social, political and environmental change' (Adger 2000). Such adaptation to change can be anticipatory (in preparation for a predictable change) and reactive (in response to a past or ongoing disturbance). The human ability to learn facilitates a kind of resilience that—different from resilience in its technical sense—allows the system to change qualitatively. That is, the system can evolve to a new (and better) state, rather than simply return to its pre-disturbance state (Wilson 2012; Brown 2014). Adaptive capacity is generally understood to depend on factors such as the recognition of the need and willingness to adapt as well as availability of resources and the ability to deploy them appropriately (Brown and Westaway 2011).

In ageing studies, the widely adopted 'ecological model of ageing' (Lawton and Nahemow 1973) assumes that the dynamics between people and their environment are fundamental to the individual experience of the ageing process. The model relates the concepts of 'personal competence' (such as personality or access to resources) and 'environmental press' (the contextual demands of the environment, such as amenities or relationship with family and neighbours). It assumes that human adaptive behaviour is shaped simultaneously by levels of environmental press and personal competence. For instance, older people with higher levels of personal competence are expected to be in a better position to use environmental resources if met with strong environmental press (Lawton 1985; Smith 2009).

At the core of this conceptualisation are theories of adaptation and the notion of coping (Lazarus 1999). Coping is defined as 'the constantly changing cognitive and behavioural efforts to manage specific external and/or internal demands that are appraised as taxing or exceeding the resources of the person' (Lazarus and Folkman 1984). Coping can be problem-focused or emotion-focused. Problem-focused (active) coping seeks to solve problems practically. Problem-focused coping mechanisms include, for instance, finding support from neighbours. To cope actively (i.e., to resolve the source of the problem) individuals will need access to resources, as well as the ability to appraise the problem effectively (Lazarus 1993). Emotion-focused (regulative) coping works to resolve problems emotionally and prevails where resources are scarce and the problem is appraised as being outside the individual's capacity to change (Lazarus 1993). The process includes 'cognitive reappraisal (a form of cognitive change that involves constructing a potentially new reality) and emotional disclosure (an act involving the expression of strong emotions by talking or writing about negative events)' (Conkova et al. 2019).

Resilience (the capacity to face adversity, remain well and even thrive) is thought as an outcome of coping (Ryff et al. 1998). Older people are frequently faced with physical losses (such as the loss of ability and mobility) as well as practical losses (such as roles and routines) and psycho-social losses (such as the loss of family and friends, opportunities, independence and others) (Tanner 2020). Resilience comprises the skills, resources and learning that takes place in the process of responding to adversity, such as maintaining psychosocial function after the death of a spouse (McCrae and Costa Jr 1988). Resilience can therefore be defined as the process of successful management of a significant adversity using assets to counter its impact (Windle 2011). Such assets may include internal traits and external factors (G. E. Richardson 2002). Resilience itself can be an asset or resource that can be used under similar conditions in the future (Igarashi and Aldwin 2012). Different from the more technical associations with resilience as bouncing back to 'normal' after a disturbance, in older age resilience denotes the ability to continue in the face of disruptions (such as chronic illness or poverty) (Richardson and Chew-Graham 2016).

In what follows, I offer a detailed account of the ways in which older people in Bulgaria's cities and villages adapt to cope with the ever-new disruptions and disturbances they face in their everyday today. First, I present the patterns of everyday life and how they are interlinked with the resources and competences of older people; I then move on to discuss the contextual demands of their environment and how older people in Sofia and the Village adapt to address shifting challenges.

## Patterns of Everyday Life

On a typical day, I get up in the morning at 6am, sometimes at 8am. Then it's time for my husband's breakfast, my own breakfast, the hygiene of the cat. This takes about two hours. I go to the market, I have to be there by 10. Not so much because I have to buy things, but for the physical activity. I try and get some movement in. I am back at 11–11.30, I make lunch for my husband, lunch for myself, I tidy up and then by 1pm I am in bed. I solve crossword puzzles and play Sudoku. Then I get up again, I make something to eat, we have dinner, and I watch TV. I watch the news and Turkish films. War movies and Turkish movies. (Tanya, 76, June 2016)

I go shopping in the morning, there are quite a few options nearby. Sometimes I go a little further [...]. I cook, I try to help my daughter in this way. Sometimes I meet with my friends in the afternoon. We watch TV in the evenings—just like all other old people. (Iliyana, 65, July 2016)

**Fig. 3.1**  A morning gathering of elderly men in the communal park, part of their socialist-era housing complex. (Photograph: Iossifova, March 2016)

In the city, pensioners in good health usually shop in the mornings. Sometimes they arrange to meet friends outside the home, at a park (see Fig. 3.1) or coffee shop. Parks are important meeting spaces. Yovka (68), who lives in Sofia with her adult son after spending much of her working life in different parts of Bulgaria, explains:

> I am social with a lot of people in the neighbourhood. We meet here in the little park. I live in a building with mixed residents—old, young. I try not to disturb anyone. So I come here to the park, where I can meet like-minded people. (Yovka, 68, July 2016)

A coping strategy to address increasing neighbourhood turnover and alienation, rather than knocking on their neighbours' doors and feeling as if they are intruding, like Yovka, older people go to parks with the intention to meet others 'accidentally'. Coffee shops take on a similar role and are often seen as the only possibility to have contact with others. Local

**Fig. 3.2** One of Sofia's local pensioners' clubs (right). The path to the entrance is covered in snow. (Photograph: Iossifova, January 2016)

pensioners' clubs (see Fig. 3.2) are also frequent meeting points and offer a range of activities to help structure the day, such as exercise in the mornings or set times for playing cards.

As of 2012, the capital had only three homes for the elderly and one home for the elderly with dementia; however, across the 24 administrative areas of the city were strewn 21 clubs for the handicapped and pensioners (citybuild.bg 2012). These clubs had taken the place of the church once the communists had taken power in 1949 and aimed at providing a spiritual home for older generations (Kulcsár and Brădățan 2014). They are still funded by the local government and usually run by a dedicated social worker. The individuals of this description that I met during my fieldwork were all driven and devoted, going far beyond what would be expected. For instance, they regularly baked or cooked for the pensioners in their care, visited their homes after work to check on them if they had failed to show up for regular activities, or committed their own financial, emotional or physical resources to do up the usually ramshackle spaces their clubs occupied or inflict positive change through other means. However, financial difficulties and challenges to securing sufficient resources are putting the future of these critical spaces in question.

Local opinions on pensioners' clubs among older people differ. The running joke is that every bench in front of a residential building is a pensioners' club—so the more of them there are, the bigger the pensioners' club. Formally organised clubs, some argue, are therefore not needed. However, pensioners' clubs present an important part of everyday life for many others, particularly those with adult children and grandchildren abroad or who would not, otherwise, afford to take part in commercial/ paid for activities on offer elsewhere.

> Three times per week I exercise at the pensioner club. One day a week I go to the dance with live music. [...] I meet friends rarely, primarily after meetings at the club. (Ana-Maria, 77, July 2016)

In the city, albeit not necessarily co-habiting with adult children and grandchildren, many older people have family at arm's length, often even the same neighbourhood, so they can count on support when needed. Generally—and if they can afford it—they also have a greater choice of cultural activities, making them less dependent on pensioners' clubs for entertainment.

> I know there is a pensioners' club around here, but I don't go. Not because I am against these, but I don't have the time. And overall, in the provincial cities and towns they get together more. Here it's different, we have more choice, there's more diversity. (Yana, 74, July 2016)

Whilst older people in good health are quite positive about their everyday lives, days are perceived as particularly monotonous by those who are bedbound or cannot leave the house for other reasons. Many describe how much their daily lives differ depending on how well they feel healthwise, days turning monotonous and depressing if in ill health:

> A typical day? There are two parallel worlds. If I feel good, I get up in the morning and I listen to music. I don't listen to the news. Sometimes I can sleep very late. I don't like home work. I like it to be clean and tidy, but I don't like doing it myself. Tuesdays we meet, us elderly ladies, they are friends of a friend of mine. I've got a friend who's got problems with her hips so I go and pick her up. We go to a coffee shop. We eat popcorn. Not

everyone can spend a lot of money. Then I come home. I eat lunch—3/4 times a week my brother will be here so I am not all alone. I have a nap. I listen to music, I read, I watch some serial. I go out sometimes with my friend. But when I am not well, it's as if the sun is gone. I have to take so much medication. I'm like a walking encyclopaedia of illnesses. The heart, the kidneys ... You name it. (Violeta, 80, July 2016)

A typical day nowadays—I get up, I eat a little so that I can take my medication, there are days when I can't go outside, so I lie in bed, and that's it. [...] I spend most of my time here, in my bedroom, in my bed. I spend so much time at home these days, I can feel I'm old. (Milka, 78, July 2016)

I have more than enough time and I have nothing to do. I am not well in terms of health, so I spend most of my time lying around. I have difficulty walking. So I try to do all shopping in one go. We have a small store nearby, although they don't have a buffet, where I could buy something ready-made. I read a lot, although that's getting increasingly difficult. Sometimes a friend or the other comes to visit. My days are very normal. Very average. It's not nice to have so much time. Especially for someone who has worked all their life and spent time with people, it's difficult to have too much time. But there's nothing one can do. One gets used to that. I see the children sometimes. (Albena, 86, July 2016)

Time is organised around the needs of others where older people have taken on caring responsibilities, such as looking after pets, grandchildren or adult relatives in need of support. This may include that grandchildren are picked up from nursery or school in the afternoon to spend the rest of the day with their grandparents until their parents collect them. Or it may involve travelling to the grandchild's home to stay with them during the day. Sometimes older people help with other activities, too, such as cooking or cleaning their adult children's home.

A typical day in Sofia is very monotonous. You get up, you tidy up, you shop, and you cook. In the countryside, you go outside, you spend a lot of time outside. This is very important for health. (Lyudmila, 86, July 2016)

Like Lyudmila, many older people live lives between the city and the village and usually find their daily lives in the countryside more fulfilling because they are not confined to the home and can enjoy what they consider a healthier environment. I will discuss this further in Chap. 4.

The daily lives of older people in the Village are typically structured around work in the field or garden. For example, Rositsa (82), who returned to the Village in 1990 after 35 years in Sofia, still enjoys having the possibility to work her garden:

> I get up at 5am on a typical day. I start work outside until about 8.30, then at 9am I have breakfast, I continue until 11am with work outside. I am no longer as organised as I used to be. Sometimes I sit down for a little bit. But I always have a plan. Then I have lunch. Around 2pm I watch a movie on TV. That's my break. I watch the romantic German movies. Then in the afternoon I'm on the phone, or I have some social obligation. My days are full. In the evenings—6.30 or so, if nobody comes to visit, I go to see people. Find out who is sick, go visit, see if I can help. It depends when I get home, usually around 8.30 I'm back. I watch the news. I go to bed around 10.30 or 11.30 pm. (Rositsa, 82, July 2016)

> I get up at 6am. I have breakfast. I wash, I shave. I shave every other day. I go to the square. I have coffee every morning at the town square. [...] We meet there at 9am and stay until 10, sometimes longer. It's men only. We discuss all rumours. Who has died, who has stolen... It's seven-eight of us. We have our coffee. We go for a walk. At 11 or 11.30 I am back at home ready for lunch. My daughter-in-law cooks and puts my lunch in the fridge. Ready to be heated up. Then I lie down with the newspaper and rest. [...] In the afternoon, I get up at 4-5 again, I go to the garden and start looking after the tomatoes, the cucumbers, all that. I don't go out in the evenings. Sometimes I go to the Pensioners Club. (Lyubomir, 87, July 2016; see Fig. 3.3)

Coffee hour with friends and neighbours around nine in the morning is very important. In contrast to the city, where older (albeit mostly well-to-do) women have successfully claimed the coffee shops and similar spaces for meetings and activities, coffee hour in the village follows clear divisions along gender lines: men meet in the coffee shops on the town square, women in their homes or gardens. The shared morning routine can be interpreted as another coping strategy enacted to meet with and participate in the everyday lives of others, as well as a kind of 'head count' designed to ensure neighbours and friends are healthy and well. It is a routine part of the everyday not just for those participating in the practice, but also for others, onlookers such as children, who—as in the quote

**Fig. 3.3** Lyubomir's garden at the edge of the Village. (Photograph: Iossifova, July 2016)

below—experience the presence of older women in a specific place and time as part of their accustomed socio-spatial environment:

> There is a child that always walks by and says: 'the grandmothers are drinking coffee again'. And when we're not here he goes: 'the grandmothers who are drinking coffee are not there'! (Nedelya, 69, July 2016)

In the Village, many of the advantages of urban life are not present. The deterioration or even vanishing of pivotal and secondary infrastructures—such as the pharmacy or cinema, respectively—complicate the experience of everyday life where most younger people have long left for better lives elsewhere. The departure of younger generations may have, as Conkova et al. (2019) note, 'fractured intergenerational solidarity; here a tradition of senior gatherings within the neighbourhood is seen as an essential element to reduce somewhat the feelings of loneliness and diversify everyday life'. Older people in the Village are therefore, indeed, ardently taking advantage of the variety of courses and gatherings on offer by the local community centre, the pensioners' club, the Russophile club, the anti-fascist union or a small range of political organisations—most notably, the Bulgarian Socialist Party (BSP)—on a daily basis.

The local BSP party secretary—82 years old Rositsa—organises meetings and events, campaigns for the construction of new playgrounds and helps resolving disputes between villagers. She looks after the sick and lonely (regardless of their membership status) and actively seeks to recruit new members. Her efforts are appreciated, and the offer of shared activities to fill their days is taken up enthusiastically by older villagers, whether or not they are party members:

> I am not a member of the party, but I sympathise. So I go to the meetings of the BSP here in town. We also go hiking [...] We've been hiking together for years. (Lyubomir, 87, July 2016)

Left behind to cope with crumbling infrastructures and sometimes crippling loneliness, older people in the Village feel the need to maintain or re-validate the value, norm and belief systems of the past and their former political identities. As everything around them changes, they experience it as quite natural to stick with 'the party' that they have known and supported for most of their lives and which, at least in some of their narratives, had given them a sense of safety and security when they were younger. This is captured by Nikolina, a widow whose only child now lives abroad and who, aged 72, feels quite lost in the emptying Village and all alone in her family's home of more than half a century:

> Everything is so strange, so not normal. I used to be BKP, then BSP, I'm too old to change. I feel loyal to this party. (Nikolina, 72, July 2016)

The Village's local community centre was established in the middle of the nineteenth century and is one of Bulgaria's oldest *chitalishta*, the regional centres which came about during the Ottoman Empire with the purpose of promoting Bulgarian culture through books, theatre, cultural activities and events. Until the Transition, the Village's *chitalishte* was well known beyond the boundaries of the village for the high quality of its activities. Still today, the centre features a library, a gallery, a women's club, children's theatre group, folk dance club, reading club, cinema, photography club and even English tutorials for students.

Like elsewhere, the local pensioners club in the Village is led by a social worker who organises trips, activities, special events and entertainment. The membership fee is 5BGN per year. According to the narratives of Village residents, the institution is largely frequented by Karakachan men. Every day between five and eight in the evening, around 25 of them come together to play cards and Backgammon. Among residents, the place is known as a drinking venue.

> I am a member of the community centre, and of the Pensioners Club. It's 5BGN membership fee per year. What do they do there? They play cards and I watch them. Women don't go there. We have someone who is looking after the club. We also have to pay for the cards when we play. They close at 8. (Lyubomir, 87, July 2016)

In the city as in the village, in the evenings older people often look forward to telephone or skype calls with children, grandchildren and great-grandchildren, who, in many cases, are spread around the world. Sometimes, friends and neighbours visit in the late afternoons and early evenings. Going out—to the cinema or theatre, like the used to do habitually in the past—is considered a luxury and reserved for special occasions; today, the main source of entertainment is watching TV.

The TV is the main source of information and, more often than not, a trusted companion. In older people's homes, especially where they live alone, TVs run night and day to provide entertainment as well as, simply put, noise, which in turn gives an illusion of others being there, present (Fig. 3.4). The success of Turkish, Indian and Russian serials that fill the afternoon and evening programmes across a range of channels is rooted in older people's need for noise. Many have installed receivers to allow them to connect to foreign broadcasters in a bid to expand their perspectives beyond the news flow in Bulgaria. But increasingly frustrated with politics, most older people steer away from news or political shows.

As these accounts demonstrate, patterns in the everyday life of older people in the city and the village are very similar. Among other important factors, such as whether or not they have caring responsibilities (for grandchildren, ill spouses or pets), to a high degree these patterns depend on the personal competences of older people, that is, their physical and

**Fig. 3.4**  Portraits—the television set in the homes of older people in Bulgaria. (Photographs: Balabanska, 2016)

mental health and the material, financial and social resources that they can access. I turn to these domains in the following sections.

## MANAGING MEAGRE PENSIONS

The retirement in Bulgaria is currently 63 years and 10 months for men and 60 years and 10 months for women; those over 70 receive additional stipends if they do not reach the guaranteed minimum pension income (Pitheckoff 2017). The average monthly pension in 2018 was 364.32 BGN (app. 162.55 GBP) (NSSI 2020). The proportion of people above the age of 65 at risk of poverty has been decreasing steadily over the last decades. Although zig-zagging in recent years, the decrease from 73.7% in 2006 to 48.9% in 2017 is notable; however, this is still substantially higher than the European average of 26.1% (eurostat 2017).

Ranging anywhere from 200 to 800 BGN, the average pension of the older urbanites who participated in this research stood at around 370 BGN, on top of which many receive disability or other allowances as well as additional income from work or property. Older people say their basic pension may be enough to pay for food or to cover their expenses for ten days—but certainly not for a month. Many rely on remittances from relatives abroad or support from members of their extended family to cover the costs of energy or medical emergencies. There are, of course, others who admit that the pension is just a lovely extra that they hardly notice on top of their main income (mostly from rent). For example, Ognian (85) worked as a lawyer and journalist before completing a PhD and starting work for the Bulgarian Academy of Sciences and teaching at the university, retiring around the year 2000:

> I receive 360 BGN. My wife is a painter and receives 160 BGN. That's the minimum pension you can get. They restituted a building in Burgas, which we now let, and this is where our income comes from. It is not much, but it is enough for us to survive. A family cannot live on 500 BGN. (Ognian, 85, September 2016)

Needless to say, on this kind of meagre pensions, older people have to prioritise how they spend their money. In the city, energy, medication and food top of the list of monthly expenses. Heating is a major expense dominating the winter months, from September to April (I will discuss this in more detail in the following section). Year-round, older people easily spend a quarter or more of their pension on life-sustaining medicines. Expenditure on food and groceries varies greatly according to individual circumstances. Strategies to make do on low pensions include hunting for food items on sale at big chain supermarkets and collecting coupons that come with the daily mail; for many, 'doing the rounds' and visiting several stores to find what they need at a price they can afford has become a daily routine. They use their monthly travel cards to get to markets and malls strewn across the city in order to source what they need at a bargain price.

> My travel card allows me to travel in the city. I go to the market—but it's very expensive here [at the nearby market], so I take the bus and go to the [other] market, it's only three stops. (Blagun, 66, July 2016)

Clothing and shoes are not a priority and often purchased at second-hand shops where they are sold by the kilo. Cultural activities—available and affordable to the 'masses' as well as encouraged and supported under state socialism—are now considered luxury and sorely missed by many older people. They sink to the bottom of the list, particularly where older people have to make a choice between spending money for their own enjoyment and supporting adult children and grandchildren. Critically, income levels delimit the type and frequency of interaction with grandchildren:

> I can't go to the opera—it's 22 BGN. I can't afford it. We used to go twice a month when we were younger. Look at us now. [...]. Oh well. Now we have to go to Pazardzhik [a town two hours' drive from Sofia] with the grandson to see some dinosaurs [referring to a Jurassic Park Entertainment Centre] (Boyana, 67, June 2016)

> So, I mentioned my grandson—imagine, nowadays spending a day with him will set me back 100 BGN. He's nine years old now. That's half a pension! (Milka, 78, July 2016)

Conditions are similar in the Village, where pensions typically range between 200 and 350 BGN. Some older people are supported by children and relatives and do not have to worry. But they are few. When pensions arrive, older people typically pay their taxes and utility bills (electricity, phone, water, TV, phone), buy their medication (10 to 80 BGN/month) and put whatever (if anything) is left away to pay for firewood during the winter months. Expenses for food and other essentials, including transportation, are kept to the bare minimum. Income from the rent of lands—quite commonly inherited and restituted land that they received after the Transition—is a welcome addition and often the only way for older people to make ends meet. Because of land consolidation processes, many do not even know where their fields are located, but regard them as an important asset and are fundamentally opposed to let these lands lie fallow. They lease to companies or corporations who generate yield for export to European Union countries.

However, the majority struggle to get by financially. Many rely on their relationships of trust with and the goodwill of local shop owners to make ends meet:

> You know, the notebooks of the shopkeepers are full with notes on people who have to come back with the money for stuff they've already given them. There are two-three people in the village who are always asking for a loan. They go, they shop, but they don't have the money to pay. They wait for their pension, they go and pay back the money they owe, and then they have to do the same thing again because they are out of money. So they start looking for loans. (Tsvetanka, 85, July 2016)

Rositsa (82), who sold the family's fields decades ago in order to help her adult children find their feet, copes in keeping a detailed balance sheet where she tracks every penny (Fig. 3.5). She spends more than 40% of her 235 BGN pension on medication, followed by water (12%) and electricity (9%). Like her, many have subscribed to the food delivery service that a local restaurant offers to older people. It sets them back 10 BGN or so per week, but provides them with some diversity and saves them from having to cook, especially if living alone:

> These are my expenses, this is how I have to count every penny. Taxes, fees, electricity, water, phone, TV, medication, and food. I get food from a local restaurant. They cook and bring food to the older people in the village. My main reason for signing up was to make sure that I eat every day; that I eat at all. It is easier that way, I don't have to go to the shop. And it is cheaper. [...] Two BGN per meal. I pay between 10 and 11 BGN per week. [...] I write down every penny I spend. I have never had to do that before. (Rositsa, 82, July 2016)

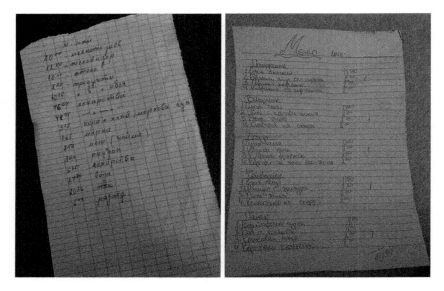

**Fig. 3.5** Left: Rositsa's balance sheet, detailing her every expense in May 2016. For instance, 10.00 BGN milk; 12.20 BGN TV subscription; 8.20 BGN groceries; 46.06 BGN medication; 41.95 BGN medication; 3.15 BGN filo pastry and bread; 27.70 BGN water; 20.36 BGN electricity. Right: the menu of the private food delivery service, offering four meal options for every day of the week, with meals costing between 0.80 and 2.90 BGN. (Photographs: Iossifova, July 2016)

## SUSTAINING AILING BODIES

Across the country, financial resources determine whether or not, and to what extent, older people—arguably the age group needing and benefitting most from functioning health care services—can access health care services. They spend a high proportion of their monthly pension on medication. A visit at the clinic for examination can cost as little as 5 BGN or as much as 50–80 BGN. Not cheap, as many remark, warning that even though it may not sound like much, it is still a considerable burden for some, particularly those on pensions around the 200 BGN mark.

> Our pensions are small. For the elderly, health care is free. If you go to the hospital, you have to pay a minimal daily fee. It used to be 3BGN, now it's 5BGN. But there are people who cannot afford even that, do you understand? Can you imagine? This is horrendous. (Violeta, 80, July 2016)

Whilst health care is generally free for pensioners, hospitalisation can be expensive as some services are available only with patient contributions. To cope, many resort to self-diagnosis using the internet, where they also identify treatment options. If 20–40 BGN for the extraction of an aching tooth, for instance, is not something they can afford, older people draw the tooth themselves to save the cost of treatment.

I am healthy at the moment. But I have to say that with democracy, they abolished health care and the health care system. I only go to see a doctor in emergency situations. Sometimes I need to see a dentist, but it is quite unaffordable. Looking after your teeth is a luxury. (Iliyana, 65, July 2016)

Bulgaria's health system was funded by public resources prior to 1989. Since then, the country has transitioned to a (mixed) public-private health care financing system and citizens rely on social health insurance (SHI) and voluntary health insurance (VHI) to access health care services. The former is administered by the National Health Insurance Fund (NHIF), the latter by for-profit companies. Insurance is compulsory and covers medication and diagnostic, treatment and rehabilitation services. VHI, covering a mere 2.4% of the population in 2013, provides faster access and certain extras (such as accommodation in a single room) and pays for services not provided on the SHI (Dimova 2016).

Contributions to the mandatory health system are 8% of monthly income for working age citizens, but out-of-pocket (OOP) payments for health care are required and very common—at 47.7%, Bulgaria has a very high rate of OOP spending (Dimova et al. 2018). The associated reduced accessibility of health care and services presents significant challenges and impacts already disadvantaged groups and older people in particular (WHO 2017), even though the state covers emergency care in life-threatening situations where individuals—up to 12% of the population (Dimova et al. 2018)—are not insured (Dimova 2016).

I had cancer, I was not fully hospitalised. I went to the hospital and came back home. Only for the surgery I had to be hospitalised. When I first learned I was sick, I first went to the central oncology, then to another polyclinic, and [eventually I settled on a private hospital to be treated by a friend.] It makes a difference to be welcomed with a smile. […] This therapy costs 100,000 BGN. Thus, we are paying for this—the medication—from whatever savings we had. But at least the surgery, the main treatment, is paid for by the health insurance. (Tsvetina, 73, July 2016)

I should note that the import and sale of medicine in Bulgaria has been in a persistently chaotic state. Reportedly, some 59% of imported medication is re-exported due to the logics of the market—for instance, if a drug is purchased and can be sold for EUR100 in Bulgaria but EUR200 in Germany (Balgarska Telegrafna Agenziya 2018). The re-export of medicines leads to frequently occurring shortages in life-saving medicines across Bulgarian pharmacies. One major incident was the failed delivery of life-saving drugs to pharmacies in 2005, leading to hunger strikes among dialysis and cancer patients who suffered life-threatening conditions as a consequence (Gancheva and Chengelova 2006). Bulgaria passed legislation with the aim to restrict parallel export of medicines, but this was repealed in 2015 (EIU 2017).

Today the country's health care system overall faces severe challenges owing to poor service provision, inefficient health system management and unfavourable demographic characteristics (Pavlova et al. 2017). Health care provision organisations are autonomous and provided by the private sector. Hospitalisation rates are relatively high, suggesting that different levels of care are not well coordinated or integrated and that ambulatory care services are not used to capacity (Dimova et al. 2018). It is remarkable that Bulgaria is on par with countries like Germany when it comes to indicators such as the number of practicing medical doctors per 100,000 residents—425 in 2017, well above the United Kingdom (281) (eurostat 2020b). That said, the number of general practitioners (GPs) has been declining, affecting rural and remote areas disproportionally.

The Village is no exception. In fact, it is quite representative of the trends in villages across the country. Older people's accounts usually revolve around comparisons between health care service provision now and prior to 1989:

> Look, they don't tell the history of the past, of socialism. It wasn't all bad. We used to have everything. The fridge was full. Now, nothing. We don't even have a pharmacy. (Snezhana, 74, July 2016)

For instance, they recall that there used to be a pharmacy in the Village—and now there is none. Even when it was still there, some older people were avoiding to collect their medication there as it did not offer the discounts to which they are legally entitled. Regardless, it did not survive the Transition and had to close, making access to medication a difficult challenge for everyone in the village. To cope, older people started going to the closest city (Kazanluk) to collect their medication. Alternatively, they now rely on the considerate pharmacist in a nearby village along the Kazanluk-Village bus route. Once ordered and paid for, the pharmacist there sends medication to the Village on the bus to spare older people another trip to collect it.

Another bone of contention is the Village doctor. There still is one, but their surgery is open only on a couple of days a week. Their mandate stipulates that they should be in the village on odd dates and keep the surgery open from 8 am to 1 pm. But people report that they are usually present only between 9.30 and 11 am and that their waiting room is so cold that one can be sure to catch something there and get sick even if healthy to begin with. Thus, many avoid seeing the village doctor. They choose a GP in Kazanluk, despite the required cumbersome bus trip.

> I have high blood pressure, and I need prescriptions for this every month, so I have to go to Kazanluk for the prescription. And for the pharmacy. I spend 40-50BGN a month for medication. And I have to pay 1BGN every time I go to see the doctor. (Nikolina, 72, July 2016)

Another common strategy of coping with the lack of affordable and reliable health care is to avoid having to see doctors or taking medication altogether. Rather, to stay healthy older people make every effort to grow their own food and apply the traditional knowledge passed down by their ancestors to collect herbs on hikes in the mountains, dry them at home and make tea and other remedies (Fig. 3.6).

## MAKING DO IN TATTERED HOMES

In the city, many live alone, sometimes in apartments that they downsized to once they lost their spouse. Not uncommonly, older people live with their adult children or grandchildren, occupying separate rooms. However, arrangements are not always perfect; for instance, due to the typical layout of housing built in the 1950s, some occupy spaces designed as living rooms to connect to adjoining bedrooms, thus lacking privacy and becoming subject to frequent disruption by their co-habitants. In this way, where space is already scarce, living conditions can be particularly challenging:

> The apartment is two rooms and a living room, but to be honest, it's very cramped. The rooms are narrow. I am not happy. [...] I have divided the

**Fig. 3.6** Lyubomir (87) still goes hiking every weekend to collect herbs, seen here drying on a newspaper in his home. (Photograph: Iossifova, July 2016)

apartment between my two sons, a room each. One son has a boiler. I don't want to enter his space. I heat my water for bathing with the small water heater. I prefer to be independent. (Yovka, 68, July 2016)

One son took one room when he got divorced, my other son took the other room. Unfortunately, to get to his room he has to go through the living room—where I live. So he passes through whenever he wants. Of course, we get along well, but it's quite annoying. (Tsvetelina, 72, July 2016; see Fig. 3.7)

Some older people live with extended family. This allows them to take care of others when they have fallen ill or because they are older and in more need of support. Formal arrangements are sometimes in place, such as taking care of an older or ill relative in exchange for rent-free living.

**Fig. 3.7** Tsvetelina's living/bedroom—the 'through-room' which her adult live-in sons have to pass to get to their bedrooms. (Photograph: Balabanska, July 2016)

I have only lived in Lozenets for a year. I live with my cousin. […] I came to be with her because she is not well. She is ill. She forgets, she falls easily. Now she seems better. […] I don't go for a lot of walks, I stay here with my cousin. So we don't go much outside. […] I clean. Her daughter has to work. She cooks and brings food. I don't eat much. Mostly fruits and vegetables. […] I don't pay rent because I look after her. […] My cousin's apartment has two bedrooms, I live in the small room, she in the other, and then there is the living room, the kitchen. Her daughter has a separate apartment, she is doing well. I don't pay for being here. The apartment is very warm, I don't even need to put on the heating. (Miroslava, 74, July 2016)

In addition to the options of living alone, with close or with extended family, there is also the commuter type: the older adult who lives independently but spends a few days a week with close family. Some very few who have a spare room can afford to live with a designated live-in caretaker For instance, 94-year-old Bogdan, who has been living in his current apartment for more than half a century. The plot belonged to his wife's father and the whole family took out a loan to build the house of which his apartment is part. When his wife was still alive, they used to spend much time at their villa in the countryside, but this is no longer of interest to Bogdan. He does not see much of his two children, four grandchildren and two great-grandchildren because they are all in different countries abroad. Due to his old age, Bogdan needs help. Someone looks after him in exchange for a roof above their head:

> My flat has two bedrooms, a living room, toilet, and bath. Probably around 100sm. I only use one room. The woman who helps me stays in the kitchen. She is my main social contact. […] She cleans, she shops for groceries, she helps me. My children don't come to visit. Nobody ever comes to visit. (Bogdan, 94, July 2016)

The situation in the Village is very similar. Those who live alone are widows or widowers whose adult children and grandchildren have moved to the city or abroad to find work. They thus live in oversized houses, designed and built to shelter a good many people and far too big for one person alone. To cope with feelings of abandonment—even fears and anxieties of living alone—many have made arrangements to spend night and day huddled up in only one of the many rooms of their house, usually that hosting their wood-burning stove. Some few live with adult children and grandchildren, but this is the exception, not the rule, because most people of working age or younger have moved to the city or abroad to find work. Where generations do share a roof, their co-habitation is often formalised through agreements that delineate each party's rights and responsibilities. There had been too many instances of older people being cheated out of their homes in the past, so most are eager to make sure this does not happen to them. Thus, in spite of usually warm and loving relationships of mutual support between older people and their next of kin, a great many have signed contractual agreements ensuring their rights to remain in their homes and be cared for in exchange for the transfer of ownership of buildings and land to their younger co-habitants.

> I have a house, upstairs is my son's family, I am here on the ground floor, but I've given the bedroom to my grandson. I sleep here [in the living room], I have a great view, and the TV. I also have a toilet on my floor. […] Upstairs they have everything, too. […] It's my house, but I've written it

over to my children, so that they have to listen to me. I have made sure they have to look after me [until I die]. You know there are people who have transferred their homes to their children or their nephews and nieces only to come home one day and realise the locks have been changed and they can't even enter their homes anymore. (Lyubomir, 92, July 2016)

Older people tend to keep animals—and especially dogs—as companions and guardians that they refer to lovingly and describe as if they were human. Tsvetanka (85), a widow who had come back to live in the Village some 25 years earlier, her only daughter living in Vienna to support her grandson, elaborates:

I live here alone with my dog. I am very calm when I am with him. Sometimes I lie in bed and I wonder: did I lock the door? Then I know it doesn't matter, because the dog is right there, at the foot of my bed. (Tsvetanka, 85, July 2016)

This is echoed by older people in the city. Just as noted by Hall (2019), 'pets were described as being family members who offered comfort, security and felt connections'. Such more-than-human relations offer 'physically, bodily and emotionally close connections' (Hall 2019) for older people in the city and Village alike:

[My son] took me to the eye doctor one day and said, mother, [what else can I do for] you? I told him—listen, don't worry about me. I can take care of myself. See, the dog needs walking... I've got things to do. On a typical day, I take the dog for a walk, I make her breakfast when we come back and I have a cup of tea and something to eat. [...] She likes peppers. [...] My dog and I, we live here in the corner. When it's cold, she knows. Her name is Ivanka. She comes and lies by my side with her back to me and that's how we keep warm. (Marina, 75, July 2016)

Heating one's home has been an escalating problem since the 1990s. Municipal heating plants were developed in tandem with apartment blocks during socialism, but their working conditions were patchy at best and served only some 20–25% of urban units. Therefore, many units continue, to this day, to rely on a variety of fuels for heating, with a large proportion of electric heating and thus high levels of vulnerability, particularly in winter (Hoffman and Koleva 1993). In 2009, for instance, in the middle of a very cold winter with temperatures of 15 degrees and more below zero, a row between Russia and the Ukraine around gas supplies and prices resulted in Bulgaria's cut-off from supply over several days. Gas shortages meant that central heating as well as a series of industrial processes had to be stopped entirely, costing the country hundreds of millions in loss (Connolly 2009). Although exporting electricity, the country serves as a transit corridor and is dependent almost entirely on oil and gas imports from Russia (Bouzarovski et al. 2012). Bulgaria tops fuel poverty statistics in the European Union, with 30.1% of households unable to keep their home adequately warm in 2019 (eurostat 2020a).

Almost half of Sofia's housing units today are part of big—prefabricated—housing complexes (containing countless *panelki,* Bulgaria's interpretation of the standardised concrete apartment block; see Fig. 3.8) which were originally centrally maintained. These complexes are suffering from faulty utilities (in some cases even absent utilities, such as central heating or land lines), structural failures and dilapidated local

**Fig.    3.8** *Panelki*—Bulgaria's    interpretation    of    prefabricated    housing. (Photograph: Iossifova, March 2016)

environments (Staddon and Mollov 2000). The lack of maintenance and upkeep can be felt in every aspect of the built environment, including the infrastructure that supports the *panelka*. Lifts, for instance, albeit vital for the functioning of building blocks of six storeys and more, are often out of order. As the municipality no longer allocates funds to the maintenance of these complexes, building stock continues to deteriorate. It is expected that all residents contribute financially to the maintenance of their housing estate; however, in times of financial hardship, homeowners can only rarely afford the necessary contributions (Vesselinov and Logan 2005).

> I spend my income on food, heating (I pay about 400BGN to heat one room only). I know why it's so expensive: I contribute to the heating of the building. So from those 400 BGN I pay 200 BGN for the heating of the building. In addition, I pay for hot water. Some neighbourhoods have gas, but I'm not sure this turns out much cheaper. So from September to April, this is where my money goes. (Stanimir, 76, July 2016)

Older people, therefore, tend to compare their current situation with their experience under state socialism, when buildings—at least according to their memories—were working as intended and they did not need worry about costs:

> If we had the money to heat it like we used to during socialism, when it was cheap, [the apartment] would be fantastic. Now we have to keep [costs] low. Now [we don't turn on the heating and] it's quite cold. We have been waiting for the retrofitting to take place. (Blagun, 66, July 2016)

With energy costs continuously on the rise, how to pay for heating is a constant worry for older people who are already having to make do with meagre pensions. Adult children often have to jump in to cover the costs of energy so that older people can stay warm. In the city, only few have the luxury of being able to choose between two or more residences, opting to spend the cold winter months in their smaller properties in order to save on energy costs. They tend to look back to a history of energy and heating transitions, ranging from the familiar petroleum oven of the 1960, gas ovens and electric panels to the fashionable air conditioner, today considered most attractive because it is cheaper than other modes of heating and also very easy to operate, saving them from having to deal with petroleum, wood, coal or other types of fuel.

The majority of older people cope with rising energy costs by refraining from heating all rooms in their apartments; instead, they limit their everyday activities to only one room which they keep warm; or they carry around small appliances (like electric heaters or blankets) to keep warm as they move between 'day room' and 'night room'. Yet, there are many instances of older people who do not heat their homes at all in order to make ends meet.

> If the weather is good, I don't heat at all, I just dress warm. If the weather is really bad, I put on the electric heater and we all huddle in one room. (Yovka, 68, July 2016)

> I live with the dog in this room. I only used the central heating the first night it got cold, to make sure the pipes are ok. Since then, I have not turned on the heating. Neither have I turned on the electric oven to heat. I don't heat. We sit in the cold. I have a coat with a hood, so that's how I keep warm. (Marina, 75, July 2016)

Finally, Bulgaria, like other Southeast European countries, has experienced the neoliberalisation of forest management with impact on the everyday lives of local people. With energy prices on the rise, firewood is becoming increasingly important as a source of energy and 'illegal' lodging is the last straw for many fuel-poor households in cities and villages alike (Petrova 2014). In the Village, the majority of older people rely on wood-burning stoves to heat their homes and cook (Fig. 3.9). Here, too, they operate within carefully set limits to prevent overstretching their budgets. Usually, they save all year to afford around 4–5 cubic metres of firewood; the current rate is 75 BGN per cubic metre. Delivery costs around 15 BGN. They hire someone for 40 BGN to cut and store the wood away—in a shed, garage or cellar.

The luckier ones do not have to heat their homes at all during the winter months. They stay with their children and grandchildren abroad or relocate temporarily to the bigger cities. There are, however, the narratives of those who cannot afford to pay for their firewood. Those who go out into the woods themselves. Who, aged 70, 80 and 90 collect their own firewood and carry it home on their backs. Mladen, for instance, aged 89 when I met him. His wife and son passed away decades ago, and he now shares a small house with his unemployed nephew. Mladen spends much of his time—year in, year out—roving the woods to pick herbs and

**Fig. 3.9**  The room with the heat source is usually the only occupied room in the house. Here, the wood-burning stove in Tsvetanka's kitchen/living room/bedroom. (Photograph: Iossifova, July 2016)

mushrooms and deadwood. He ties fallen branches and sticks into bundles so that they are easier to lug home. From spring to autumn, he says, he manages to gather just about enough to get through the coldest winter months. And then, there are the stories of those others, who cannot walk or carry, and who do no longer mind throwing anything into their wood-burning oven, as long as it would burn: old newspapers, books and plastic bottles.

More recently, the state has come around to provide at least some minimum support, hardly sufficient to make a real difference, to those who have trouble paying for energy. Yet for many older people, the bureaucratic aspects of even just submitting an application are a hurdle, preventing particularly those who may need the top-ups most from seeking help—those struggling with chronic illnesses or with mobility issues:

> There is a new regulation, they give you 50 BGN for five months if you can prove that you wouldn't be able to pay for [your heating] otherwise. But you need to bring so many documents, it's unbelievable. [The place where you submit these documents is] very far away, and you have to go in person. I simply can't. (Miroslava, 74, July 2016)

Being unable to heat the own home due to rising energy costs, deteriorating building stock or outdated and dysfunctional heating technology emerges as an important hurdle to health and wellbeing in ageing. Physical and spatial strategies in the everyday to minimise the burden on their limited financial resources include moving in with family and friends to share costs, contracting the spaces they occupy or wearing multiple and thick layers of clothing so as not to have to heat at all.

## CONCLUSIONS

Bulgaria's older people have developed a range of adaptive mechanisms in response to old and new challenges in their daily lives. In this chapter, I have demonstrated just how remarkably resilient they are, succeeding to adapt to meagre pensions, ailing bodies and tattered homes in the face of extraordinarily uncertain times. Framing my enquiry through the human ecosystem framework (Machlis et al. 1997, see Chap. 1) has allowed me to capture at least some of the socioeconomic, natural and cultural resources entangled with the lives of older people in Bulgaria, as well as aspects of the social cycles and institutions that shape them.

The concepts of coping, adaptation and resilience support my analysis older people's everyday lives. Typical patterns depend on physical and mental health. They differ based on socioeconomic status and, subsequently, on what one can afford to do, and where. They vary according to socio-spatial and environmental factors, such as location (whether a person lives in the city or in the Village). Sofia has been continuously growing, and whilst the state has been curbing or even withdrawing public services in recent decades, urban life continues to offer advantages, such as the availability and density of facilities and amenities ranging from shops to hospitals. In contrast, like elsewhere in the countryside, the Village's population has been shrinking relentlessly in recent years, and with it the number of services, facilities and amenities on offer.

I have provided a number of oftentimes distressing accounts of the ways in which older people address the challenges resulting from insufficient resources of different kinds—that is, how they cope with taxing situations. For instance, it matters with regard to their coping mechanisms whether a person lives alone, with a spouse, an adult child or paid caregiver. I find that older people, where they live alone, address the physical and emotional fall-out of their state of abandonment through interaction with peers, pets and TVs and more sophisticated technology (which I discuss in more detail in Chap. 4). They cope with insufficient pensions by topping them up through income from rent (where this is available), post-retirement jobs, support from children and grandchildren and remittances from relatives abroad. They resort to do-it-yourself solutions where they have to cope with the crumbling or even complete lack of (affordable) health care infrastructure. Often unable to cover the cost of heating, they collect deadwood or abstain from heating at all.

Overall, older people have adapted or are in the process of adapting remarkably well to all kinds of disturbances to the human ecosystems of which they are part. Some of the narratives presented here, however, demonstrate that even though they may appear incredibly resilient under conditions of uncertainty, older people's lives continue to be marked by hardship and struggle. As Lesutis (2019) notes: 'overemphasising everyday coping as resistance can end up depoliticising structural inequalities' (see also Ferguson and Harman 2015). And as I show extensively in this and the previous chapters, structural inequalities in contemporary Bulgaria persist and amplify as the nation's population ages. In Chap. 4, I move to identify the role of place in the experience of ageing in Bulgaria, particularly transforming relationships with place across rural-urban and international divides.

# REFERENCES

Adger, W. N. (2000). Social and ecological resilience: Are they related? *Progress in Human Geography, 24*(3), 347–364.

Balgarska Telegrafna Agenziya. (2018). *Patients' organizations: 59% of imported medicines in Bulgaria re-exported later on.* Sofia: Balgarska Telegrafna Agenziya.

Bouzarovski, S., Petrova, S., & Sarlamanov, R. (2012). Energy poverty policies in the EU: A critical perspective. *Energy Policy, 49*(Supplement C), 76–82. https://doi.org/10.1016/j.enpol.2012.01.033.

Brown, K. (2014). Global environmental change I: A social turn for resilience? *Progress in Human Geography, 38*(1), 107–117.

Brown, K., & Westaway, E. (2011). Agency, capacity, and resilience to environmental change: Lessons from human development, well-being, and disasters. *Annual Review of Environment and Resources, 36*(1), 321–342. https://doi.org/10.1146/annurev-environ-052610-092905.

citybuild.bg. (2012). *There will be 21 clubs for pensioners in Sofia [ОБЩО 21 КЛУБОВЕ НА ПЕНСИОНЕРА ЩЕ ИМА В СОФИЯ]. citybuild.bg.*

Conkova, N., Vullnetari, J., King, R., & Fokkema, T. (2019). "Left like stones in the middle of the road": Narratives of aging alone and coping strategies in rural Albania and Bulgaria. *The Journals of Gerontology: Series B, 74*(8), 1492–1500.

Connolly, K. (2009, 10 January). Big chill: Bulgarians battle to keep warm. The Guardian. Retrieved from https://www.theguardian.com/business/2009/jan/10/gas-dispute-bulgaria-russiagazprom.

Dimova, A. (2016). Bulgaria. In A. Sagan & S. Thomson (Eds.), *Voluntary health insurance in Europe: Country experience* (Vol. 42). Copenhagen: European Observatory on Health Systems and Policies.

Dimova, A., Rohova, M., Koeva, S., Atanasova, E., Koeva-Dimitrova, L., Kostadinova, T., et al. (2018). *Bulgaria: Health system review* (Health Systems in Transition) (Vol. 20). Copenhagen: World Health Organization. Regional Office for Europe.

EIU. (2017). *Cancer medicines shortages in Europe: Policy recommendations to prevent and manage shortages.* London, New York, Hong Kong: The Economist Intelligence Unit (Healthcare).

eurostat. (2017). *ECHI data tool.* Brussels.

eurostat. (2020a). Population unable to keep home adequately warm by poverty status. In eurostat (Ed.).

eurostat. (2020b). *Practising physicians.* Brussels: eurostat.

Ferguson, L., & Harman, S. (2015). Gender and infrastructure in the World Bank. *Development Policy Review, 33*(5), 653–671. https://doi.org/10.1111/dpr.12128.

Gancheva, V., & Chengelova, E. (2006). Social Services for the Elderly [*Socialni Uslugi za Horata ot Tretata Vuzrast*]. Retrieved from http://www.omda.bg/public/biblioteka/vyara_gancheva/vyara_emi/vyara_emilia_00.htm.

Hall, S. M. (2019). *Everyday life in austerity: Family, friends and intimate relations*. Cham: Palgrave Macmillan.

Hoffman, M. L., & Koleva, M. T. (1993). Housing policy reform in Bulgaria. *Cities, 10*(3), 208–223. https://doi.org/10.1016/0264-2751(93)90031-D.

Igarashi, H., & Aldwin, C. (2012). An ecological model of resilience in late life. *Annual Review of Gerontology and Geriatrics, 32*(1), 115–130.

Kulcsár, L. J., & Brădățan, C. (2014). The greying periphery—Ageing and community development in rural Romania and Bulgaria. *Europe-Asia Studies, 66*(5), 794–810. https://doi.org/10.1080/09668136.2014.886861.

Lawton, M. P. (1985). The elderly in context: Perspectives from environmental psychology and gerontology. *Environment and Behavior, 17*(4), 501–519.

Lawton, M. P., & Nahemow, L. (1973). Ecology and the aging process. In C. Eisdorfer & M. P. Lawton (Eds.), *Psychology of adult development and aging* (pp. 619–624). Washington, DC: American Psychology Association.

Lazarus, R. S. (1993). Coping theory and research: Past, present, and future. *Psychosomatic Medicine, 55*(3), 234–247. https://doi.org/10.1097/00006842-199305000-00002.

Lazarus, R. S. (1999). *Stress and emotion: A new synthesis*. New York: Springer Publishing Company.

Lazarus, R. S., & Folkman, S. (1984). *Stress, appraisal, and coping*. New York: Springer Publishing.

Lesutis, G. (2019). The non-politics of abandonment: Resource extractivisim, precarity and coping in Tete, Mozambique. *Political Geography, 72*, 43–51. https://doi.org/10.1016/j.polgeo.2019.03.007.

Machlis, G. E., Force, J. E., & Burch, W. R. (1997). The human ecosystem part I: The human ecosystem as an organizing concept in ecosystem management. *Society & Natural Resources: An International Journal, 10*(4), 347–367.

McCrae, R. R., & Costa, P. T., Jr. (1988). Psychological resilience among widowed men and women: A 10-year follow-up of a national sample. *Journal of Social Issues, 44*(3), 129–142.

NSSI. (2020). *Average number of pensioners and pension 2012–2018*.

Pavlova, M., Atanasova, E., Moutafova, E., Sowa, A., Kowalska-Bobko, I., Domagala, A., et al. (2017). Political will against funds deficiency: Health promotion for older people in Bulgaria. *Zeszyty Naukowe Ochrony Zdrowia. Zdrowie Publiczne i Zarzadzanie, 15*(1), 108.

Petrova, S. (2014). Contesting forest neoliberalization: Recombinant geographies of 'illegal' logging in the Balkans. *Geoforum, 55*(Supplement C), 13–21. https://doi.org/10.1016/j.geoforum.2014.04.008.

Pitheckoff, N. (2017). Aging in the Republic of Bulgaria. *The Gerontologist, 57*(5), 809–815. https://doi.org/10.1093/geront/gnx075.

Richardson, G. E. (2002). The metatheory of resilience and resiliency. *Journal of Clinical Psychology, 58*(3), 307–321.

Richardson, J. C., & Chew-Graham, C. A. (2016). Resilience and well-being. In *Mental health and older people* (pp. 9–17). Cham: Springer.

Ryff, C. D., Love, G. D., Essex, M. J., & Singer, B. (1998). Resilience in adulthood and later life. In J. Lomranz (Ed.), *Handbook of aging and mental health* (pp. 69–96). New York: Plenum Press.

Smith, A. E. (2009). *Ageing in urban neighbourhoods: Place attachment and social exclusion.* Bristol: Policy Press.

Staddon, C., & Mollov, B. (2000). City profile: Sofia, Bulgaria. *Cities, 17*(5), 379–387. https://doi.org/10.1016/S0264-2751(00)00037-8.

Tanner, D. (2020). Resilience and older people. In N. Thompson & G. R. Cox (Eds.), *Promoting resilience: Responding to adversity, vulnerability and loss* (pp. 121–126). New York and London: Routledge.

Vesselinov, E., & Logan, J. R. (2005). Mixed success: Economic stability and urban inequality in Sofia. In F. I. Hamilton, K. D. Andrews, & N. Pichler-Milanović (Eds.), *Transformation of cities in central and Eastern Europe: Towards globalization* (pp. 364–398). Tokyo: United Nations University Press.

WHO. (2017). *Bulgaria: Highlights on health and well-being.* Copenhagen: WHO Regional Office for Europe.

Wilson, G. (2012). *Community resilience and environmental transitions.* Oxon: Routledge.

Windle, G. (2011). What is resilience? A review and concept analysis. *Reviews in Clinical Gerontology, 21*(2), 152–169.

# CHAPTER 4

# Dislocating 'Ageing in Place': From Multi-local to Transnational

**Abstract** Iossifova critically reflects on the notion and agenda of 'ageing in place' to argue that it is inherently inapt to capture or address the main concerns of older people in Bulgaria. She presents how older people navigate and experience the transformation of their neighbourhoods in Bulgaria's capital and the Village in the Bulgarian Balkans. She argues that Bulgaria's rural-urban reciprocity has fostered profoundly translocal ageing processes, whereby older people oscillate between homes and social networks. She shows that ageing is enmeshed with recent trends in migration, producing a kind of transnational ageing utterly incompatible with 'ageing in place' agendas. Iossifova argues that translocal ageing enables older people today to cope with the reality of being both abandoned by the state and left behind by adult children and grandchildren.

**Keywords** Bulgaria • Older people • 'Ageing in place' • Mobility • Migration • Multi-local ageing • Rural-urban reciprocity • Transnational ageing • Translocal ageing • G0 parenting/grandparenting

My paternal grandmother lived with my uncle and his family in Sofia during the colder months of the year. By the time she had become too old to make the trip to the Village by herself, the Transition, restitution and a series of other changes had complicated my uncle and his family's lives, making it impossible for them to permanently take her in. It was decided that she would move into my family's former apartment and share it with a young professional who, in exchange for a discount in their rent, would look after her and make sure she had everything she needed. Over the years that followed, on visits home I was surprised to find that she had, in no time at all, recreated relations my other grandmother had built over the 40-odd years she had lived in this apartment. She had established strong networks of mutual support with neighbours, young and old, and had made the neighbourhood hers. The experience of socioeconomic and political shifts, the rural-urban reciprocity and associated frequent back-and forth and—in the case of my grandmother—the years spent abroad at young age had all contributed to her ability to quickly and seemingly effortlessly adapt to new living arrangements and find her place within an otherwise well-established community. Her resilience proved extraordinary.

*    *    *

In the Western context, it has been assumed for a long time that for older people, everyday life is like 'long-running improvisational theatre in which the settings, characters and actions are familiar and in which the changes are mostly in the forms of new episodes rather than entirely new plays' (Atchley 1971). In this chapter, I argue that in the case of Bulgaria (and possibly the wider Global East ([Müller 2020]), this assumption can no longer be upheld. In the everyday lives of older people in Bulgaria today, changes frequently *do* come as entirely new plays. My grandmother's experience, sketched at the start of this chapter, is echoed in the narratives of many older people who participated in the research for this book.

In this chapter, as I do in Chaps. 2 and 3, I draw on interviews with older people in Sofia and in the Village in order to understand the role of place in ageing. The narratives are extracted from their responses to open-ended questions querying categories of the human ecosystem framework (Machlis et al. 1997; see Chap. 1). Although entangled with a range of other categories, the ones particularly relevant in this chapter include *social cycles* (patterns of the everyday), *socioeconomic resources* (information, population), *cultural resources* (organisation) and *natural resources* (energy). I show how older people navigate familiar places in the Village and in Sofia, and how they experience the recent and fundamental socio-spatial transformation of their local environments. This leads me to question the notion of 'place' in the context of ageing altogether. I then move to examine the typically Bulgarian rural-urban reciprocity and associated oscillation between homes in the city and homes in the countryside to arrive at the conclusion that ageing in Bulgaria has long been a translocal experience. Finally, I describe the evolution of translocal ageing into transnational ageing as a consequence of international migration and allude briefly to the role of information technology in enabling close and intimate relationships in spite of great physical distances.

In the following section, I offer a critique of the notion of 'ageing in place' and the agenda attached to it. I propose that there is urgent need to develop new approaches that are closely aligned with the realities and requirements of older people ageing in the context of the Global East—a context, as I have stated before, to large parts defined by uncertainty.

## RETHINKING PLACE IN AGEING

Over the last couple of decades, successful ageing has been connected conceptually with specific aspects of space and place, particularly the paradigm that it is highly dependent on remaining in (familiar) places for as long as possible in order to secure continuity across identity, independence, social support and relationships (e.g., Wiles 2005; Peace et al. 2011). It is assumed that ageing in a familiar environment is of benefit to older people for a large variety of reasons, most importantly because it enables them to live more independent lives whilst remaining embedded within a local service network (Rowles 1993). This builds largely on the assumption that ageing people 'have the need and the tendency to maintain the same personalities, habits and perspectives that they have developed over their life course' (Estes et al. 2003) and that therefore they have the need for continuity with regard to place. It is assumed that people should 'stay in charge of their own lives for as long as possible as they age and, where possible, [...] contribute to the economy and society' (European Commission 2018).

The 'ageing in place' argument is built around the concepts of place attachment and independence. The literature links physical attachment with intimate knowledge of place developed from long-term residence in place; independence results from intimate knowledge of place which compensates for the age-related decline in health (or personal competence more generally) (Lawton 1985). For instance, older people may—as we have seen in Chap. 3, but for different reasons—choose to restrict their spatial experience in occupying only one floor of a multi-storey house, thereby increasing their feeling of control (by knowing their restricted environment better) and therefore experiencing their environment more positively (Gory et al. 1985; Francis 1989). It is argued that 'ageing in place', in a familiar environment, allows older people to cope with spatial restrictions and increase psychological wellbeing (Howden-Chapman et al. 1999; Phillipson 2004; Smith 2009). Furthermore, place is understood to provide important functions for a person's identity process, for instance in offering the opportunity for life reviews: remembering where one used to play as child, got married or sent one's children to school is said to contribute to psychological health and greater quality of life (Borglin et al. 2005).

The own home offers the advantages of familiarity and the opportunity for life review (Baker and Prince 1991). Continuing to live at home as long as possible has been presented as the most desirable option because it allows the elderly to remain independent from institutional structures and maintain self-determined ways of life—that is, to remain independent (Hillcoat-Nallétamby 2014). However, critics note that for some ageing people, living outside of the own home, in institutional settings, may bring advantages; for instance, this may make care around the clock accessible, offer the opportunity to build new relations or do activities together with others (Martin et al. 2005; Boyle 2005). The 'ageing in place' paradigm neglects the possibilities that the idea of home may fall prey to romantisation (Oldman and Quilgars 1999) or that the home may be experienced as a trap contributing to isolation (Milligan 2012). For example, Fried (2000) notes that people with fewer physical, economic or social opportunities may compensate in attaching more sense of belonging to notions of place, precisely because their need for shared conditions in life-long communities is greater. Consequently, older people may take pride in the places they inhabit even though these places might be unsuitable to accommodate their current or future needs as ageing individuals (Severinsen et al. 2016). To reconcile this with (increasingly) limited physical, economic or social opportunities, older people may thus be prone to generating narratives of place-related identity, of social bounds (with people of the present as well as the dead of the past), and of housing as part of their life histories in place.

My evaluation of the 'ageing in place' paradigm is that it is profoundly archaic. It is rooted in the 'assumption that all social relations are organized within self-enclosed, discretely bounded territorial containers'—what Brenner (2004) pointedly calls 'methodological territorialism'.

Johansson et al. (2013) write: 'In relation to emphases on person-environment fit, "ageing in place" has been critiqued for promulgating a static view of place as a pre-existing "container" existing apart from ageing persons, rather than as a process that involves continual transactions between space and ageing persons that transform both'. It has been widely (and somewhat uncritically) adopted as a key conceptual framework and even WHO policy guideline to improve the lives of older people based on experiences in countries such as the United Kingdom and Sweden (Black 2008; Means and Evans 2012).

The ontological and epistemological focus on the Global North once more neglects the wide diversity of experiences and therefore context-specific needs elsewhere, including in the Global South and East. Frameworks that build on skewed evidence must always be questioned, expanded and tweaked, especially in times of globalisation and unprecedented migration. The literature on ageing and place attachment is complicit in propagating a 'romantic ideal' (Johansson et al. 2013) of aging in place that overlooks the varying and changing relationships of older people to their homes. It is particularly negligent of the complex experiences of older people who migrate and/or are part of transnational or multi-local families (i.e., families scattered across and straddling multiple places and residences, thus instantiating the processes of globalisation and associated migration [Borell 2003]).

Globally, the last few decades have been marked by migration: unprecedented rural-to-urban migration in China, the East-West migration of young people in Europe following the fall of communism or the most recent phenomenon of refugees from Syria and other war-torn regions arriving in Europe. New mobilities are continually emerging, sometimes engendered by a wealthier band of ageing people (Botterill 2017) and theorised as consumption-led or lifestyle mobilities (Xu and Wu 2016; Benson 2011). As Johansson et al. (2013) note, 'people are travelling more than ever, emigrating, immigrating, or simply seeing the socio-cultural landscape of their neighbourhood change as people move in and out of an area'. Social, economic and cultural relations unfold across the boundaries of the family, the household, specific locales, regions and nation-states. Whilst many people may well remain dynamically embedded within their place-based context, be it at the level of the home, the neighbourhood or community (Wiles 2005; Darling 2009; Schwanen et al. 2012), there is urgent need to rethink the *boundaries of place* in view of older people's lived experiences.

'Ageing in place' does not suffice as a conceptual and analytical framework for the understanding of ageing and place relations and transactions (Cutchin et al. 2006; Cristoforetti et al. 2011). People, presumably of all ages, continuously renegotiate their relation to place. As people age, they inevitably adapt to change, including an ageing body, the need for care and a full range of new social identities, such as for instance that of the older person (Näre et al. 2017). Yet scholarship seems to confine 'thinking on ageing in place to those spaces that are geographically proximate to the ageing body or physically bounded such as the house' (Boyle et al. 2015). People with migration histories or affected by the mobility of others in their lives are particularly subjected to such adaptation needs as they negotiate their everyday lives and identities in transnational contexts. For instance, the emigration of adult children from rural Albania was found to undermine older people's self-respect, leaving them feeling like 'elderly orphans' with little choice but to follow their children abroad and to care for their grandchildren as a way of coping (King and Vullnetari 2006). Deneva (2012) illustrates on the case of Bulgarian Muslim migrants in Spain how migration disrupts the structure of care arrangements, kin expectations and family relations and the social citizenship (and therefore the opportunity to receive state welfare support) of transnational carers. This results in intergenerational and gender inequalities that remit further exploration.

Transnational ageing as a relatively recent notion is defined as the 'process of organising, shaping, and coping with life in old age in contexts which are no longer limited to the frame of a single nation state' (Horn et al. 2013). It is, of course, the result of an increasingly interconnected world, where affordable technology and infrastructure allow frequent back-and-forth movements between different countries. Although the field of transnational studies is fairly established, scholars in ageing studies have been slow to adopt a transnational perspective (Phillipson and Ahmed 2006). This has been attributed to territorial-bounded thinking, expressed, most strikingly, in the concept of 'ageing in place' which assumes 'a quasi-natural overlap between a territory and an individual's social space' (Horn and Schweppe 2017; Zhou 2012).

I support the notion of placemaking—not in the strict urban design sense (Carmona 2010), but more generally as describing the process in which people give meaning to physical spaces to create socially relevant places (Paulsen 2010). Placemaking has been suggested as an alternative conceptual approach to the relationship between ageing and place

(Johansson et al. 2013). It alludes to an active engagement with place, affording older people the agency to act and change their environment in order to adapt to change and respond to challenges (to cope). The notion reflects in a more nuanced way the complex experiences of people who are ageing in familiar and unfamiliar places and/or in-between places. The notion calls for a conceptualisation of the human-environment relationship as dynamic, fluid and complex and allowing for the consideration of more than one place as part of older people's experiences and actions, for instance, in the context of migration (Gesler and Kearns 2005). Understanding placemaking practices is therefore essential in order to create policy responses in the face of challenges arising from an ageing population. It is essential to support the resilience of older people in times of uncertainty.

In the following sections, I unpick four mobility domains that bring to the fore the diverse experiences of ageing locally and translocally, in and out of place: the navigation of familiar places, the transformation of familiar places, the oscillation between rural and urban places, and ageing transnationally.

## Navigating Familiar Places

Older people in the capital can usually rely on functioning physical and social infrastructure, including the greater density and availability of general practitioners (GPs); malls and shops to get everyday necessities; and buses, trams and trains to get around in the city. They do not find mobility to be of huge concern in the greater scheme of things. Only few hold on to their cars, usually seen as a last resort or used only for trips to the countryside. Even fewer do not use public transportation because they feel vehicles are not well maintained and mingling with others not desirable. They base their evaluation on their public transport experience during visits abroad:

> I have a friend who lives in Germany and the States, and I visit her at least once a year for 20 days. I don't like using [public transport here]. There are many mentally disturbed people on the tram, people who don't shower. It's different in Germany. There, it is my favourite. In Sofia, I walk, or my son drives me here and there. (Rada, 74, July 2016)

But apart from a very few such exceptions, most quite like—are even proud of—the state of public transportation. Polyclinics, hospitals, pharmacies, shops, community centres and family and friends are always only a short walk or tram ride away; thus the vast majority of older people walk or rely on convenient public transport. Those with mobility restrictions are quick to point out accessibility challenges, such as the height of tram steps or the lack of (functioning) lifts to get in and out of metro stations (see Fig. 4.1). But generally, older people are content with the low fares: pensioners pay only 8 BGN per month and have unlimited access to the full public transportation network in the city, meaning that even older people on the lowest of incomes can still afford to travel.

**Fig. 4.1**    Stairs leading to/from a metro station in Sofia. Lifts have been provided or are being retrofitted according to EU regulations, but many are dysfunctional, leaving people with mobility issues challenged. (Photograph: Iossifova, March 2016)

I have a pensioner's card and I use public transportation. I don't know how long they will keep the pensioners card up, it's very affordable, 8 BGN. I use the bus and the tram. I'm used to the bus. The metro is also very convenient, very fast, we're only 100 meters from the station. (Lyudmila, 86, July 2016)

Likely because their everyday lives are often structured around shopping and associated daily trips to stores, older people frequently report the appearance of shopping malls in the city. These are thought to offer conveniently everything one needs in high quality and in one place. In this context, the implementation of the metro system—a few lines over two or more decades—is perceived as a very positive development as it allows them to transverse the city in no time.

Another great improvement are the malls and the possibility to shop high-quality goods. People are so much more polite at the store. The metro has more than 30 stations, it's very well organised. (Daniel, 77, July 2016)

We have two-three buses that take us to the city centre, we have three trams which take us whenever we want; my flat is located between two metro stations. (Yana, 74, July 2016)

Within the Village, older people walk—it is a small village, 25 minutes from end to end. If they ever owned a car, they have long sold it or given it away. A large number can count on their adult children/grandchildren or younger neighbours to drive them to town when needed:

My daughter drives me if I have to go somewhere, I don't use public transportation, usually. I go to Kazanluk several times per month. We make it into an occasion, I never go alone. (Nedelya, 69, July 2016)

However, many rely on 'the bus' (there is only one) to make the (at least) monthly half-hour trip to the big city, Kazanluk, to run life-sustaining errands, such as going to the pharmacy to get their medications. The bus provides convenient service, connecting Kazanluk and the Village and stopping in two other villages along the way. It runs from seven in the morning until seven thirty at night, every thirty minutes during the week; once per hour on weekends. The return fare is 4 BGN, and although pensioners and school children receive 20% discount, it can still be a steep price to pay. Nevertheless, to many, it is a lifeline in a shifting environment (Fig. 4.2).

**Fig. 4.2** 'The square' at the centre of the Village. 'The bus' is seen waiting for passengers. (Photograph: Iossifova, July 2016)

## TRANSFORMATIONS 'IN PLACE'

So far, I have presented the main challenges that older people face and how they address them within their supposedly familiar urban and rural environments in Sofia and in the Village. I say here 'supposedly familiar' because the dramatic transitions that Bulgaria has experienced as a nation over the last few decades do not go unnoticed at the very scale of the local social and spatial environment. As already outlined in Chaps. 1 and 2, Bulgaria's recent socioeconomic and environmental transitions have led to outmigration, low fertility rates and crumbling health and care infrastructures. Most villages and cities are rapidly losing population to the capital or destinations abroad. The transition to a market economy with all its permutations and associated transformations has disrupted familiar patterns of making, maintaining and living in villages, neighbourhoods and cities. The following paragraphs are concerned with older people's experiences of shifts in their local built and social environments.

In preparation for and in response to the rapid rural-to-urban migration gripping the country between the 1940s and 1980s, the state transformed the fields, forests and villages surrounding the city core into swaths of residential complexes containing thousands of prefabricated concrete panel blocks, *panelki* (Staddon and Mollov 2000; Nikolov 2020). Neoliberal development policies following the fall of state socialism in 1989 contributed to the capital's sprawl into surrounding green areas and the development of road infrastructure of questionable quality. Most recently, Sofia was on the brink of witnessing the roll-out of the China-financed project Saint Sofia, the first corporate-led smart city project in Europe. Planned only 15 minutes from Sofia's International Airport and featuring an indoor water park (the world's largest) and prestigious golf club, among other perks (Zhuan 2017), Saint Sofia was to replace the large swathes of sunflower fields that characterise the landscape today (Milenkovic 2018). Construction, however, was delayed for several years and eventually the project fell through.

Thanks to still-existing socialist planning, in most corners of the city urban residents continue to have access to parks and gardens in walking distance. There, they look after grandchildren or meet with friends, but hardly ever neglect to mention that they used to feel safer in the past. This is usually attributed to the stricter control of public spaces under socialism; however, as they age, older people also acknowledge the loss of confidence linked with the declining functioning of their body:

> There are places to go for a walk in this neighbourhood, you can reach the [park]. But the park lost its character of a place to go for walks. I can tell you that I used to cross a dark forest between my apartment and my work place back in the days, at 11 or 12 at night, and I felt safe. I wouldn't dream of doing this nowadays. (Ognian, 85, September 2016)

Sofia's ongoing transformation does not go unnoticed by residents in neighbourhoods like Lozenets, who once considered themselves at the edge of the city and who now find that their neighbourhood is considered central and is surrounded by mushrooming blocks of housing:

It used to be all tiny houses with big gardens. This here used to be a huge yard with fruit trees, which I miss very much. There used to be a giant cherry tree, there used to be a bench underneath and a little table. But as it so happened, we had to build this ... After all, this is not too high, it's only three floors, and they didn't allow us to go too high back then. Then they started, here a storey more, there a storey more ... The environment has changed a lot. We have the metro now, they opened the little market, and they opened many small shops. It has changed, a lot. (Tsveta, 90, July 2016)

Simultaneously, older people persistently note that their neighbour-hoods have been uprooted with the moving away or passing of former neighbours and friends. They mention changes in the social make-up of urban neighbourhoods, demonstrating acute awareness of the rampant emigration of younger generations. Take Raina's observations (68), for instance. Born and bred in Sofia, she still lives in the apartment of her childhood. Her father was involved in the construction of the apartment building. These days, Raina spends her time looking after her husband who is very sick after suffering a stroke. Her daughter and grandson step in occasionally if she has to leave the city for a short while. But since her husband became sick, this hardly ever happens, leaving the family's villa in the countryside unused. Raina spends the little time she has to herself visiting with friends.

We used to meet in the park with the other neighbours, but not so much anymore. The benches got mouldy. [...] Overall, this is an ageing neigh-bourhood. I guess this is the case for most neighbourhoods, because most young people are not in Bulgaria. I am sure they will come back, eventually, the nostalgia is great. (Raina, 68, July 2016)

Things have changed a lot, and are beginning to change even faster. The small houses are disappearing, a lot of construction is happening. There are 100 or 200 new buildings around here, residential buildings, people need that. [...] I don't know where people in the neighbourhood can socialise. [...] There is a coffee shop—I thought this would be the centre for social interactions. Not at all. People live fast, people need to make money, and they don't have time to socialise. (Stefan, 65, July 2019)

Vacant apartments are quickly snapped up by much younger people, turned into offices or simply left to stand empty, withering away (see Fig 4.3).

**Fig. 4.3** One of the entrances to an apartment building in Sofia; this entrance is boarded up to block access to the building's stairwell. Many of the apartments in the building are no longer occupied. (Photograph: Iossifova, March 2016)

> The building is empty. Half have died, the other half have gone abroad. There is so much space now. We don't have small children in the building. Everyone has gone. [...] We were poor but we shared everything. [...] It used to be that someone would buzz you and they'd come in for a drink or a coffee. Now you stand in the doorway as you talk to people and you're afraid. (Blagun, July 2016)

In the Village, these sentiments somewhat reverberate with older residents who note how more and more people are moving away, going abroad and leaving their homes behind with facades crumbling and roofs giving way to snow and rain. They are weary of the handful of foreigners arriving to buy cheap land and houses and then keeping to themselves, thus changing the character of the Village and taking away its vibrancy. The Village, they say, is no longer what it used to be. And yet, they do not echo the sadness, resignation and anger described in the literature as common responses to the rapid demographic decline among older residents in rural settings (Conkova et al. 2019).

In spite of the dramatic shifts over the course of their lives (see Chap. 2) and in their day-to-day activities (see Chap. 3), older residents in the city and the Village alike adapt to their ever-changing socio-spatial environment. Likely, this has to do with their active membership in the local community and the strong networks of mutual support that they have built over time—and across space. Many divide their lives between the Village and the city; others venture even further afield and across national and even continental borders; others again rely on newly acquired technology and technological skills to connect with family and friends around the world.

## Between the City and the Village

The key elements of family in Western societies are traditionally considered to be co-residence, geographical proximity and spatial co-presence (Schier 2016). Such fundamental considerations are beginning to splinter as families are becoming increasingly diverse and spatially scattered. So-called multi-local families can be divided into two categories: those jointly practiced by a family (i.e., a family moving together between locations) and those marked by the separation of family members and the split into more than one households, geographically separated (Schier 2016). Bulgaria's older people are often part of families of the second kind. That is, they usually oscillate between households across two or more households, in the city (where they may co-reside with adult children) and in the village; or in the city/village (where their own household is located) and abroad (where their adult children/grandchildren are).

The difference between a multi-local and a transnational family is that the first notion denotes a geographical split regardless of distance, whilst the second implies a split across nation-state borders (Schier 2016). Very few studies focus on the comparison between intra- and international intergenerational relationships (e.g., on the case of Romania, see Zimmer et al. 2014). In Bulgaria, older people are often part of both. Here, I do not wish to compare the two types of relationships and their repercussions. Rather, I want to draw attention to the translocal experience of ageing in multi-local *and* transnational contexts alike as a way to help undo the fixation on the 'ageing in place' paradigm.

I should be very clear that by far not everyone in Bulgaria has a house or villa in the countryside, and that the proportion of older people in Sofia who do gradually decreases with their socioeconomic status as we transect the city from East to West. A house, a villa in the countryside is something people may have inherited from their parents and grandparents, indicating rural-to-urban migration in recent generations and speaking again of the rural-urban reciprocity tradition that Bulgaria builds on (Smollett 1985; Konstantinov and Simić 2001). Otherwise, a villa is something they may have received, bought or built under socialism—something that certainly not everyone in the city was entitled to or could afford, therefore indicating a certain degree of privilege (Parusheva and Marcheva 2010).

There is therefore the large group of people who do not have access to the luxury of translocal ageing, spent oscillating between the village and the city, the city and the village, or the village, the city and places further afield. Usually least well-off, somehow making ends meet on minimal pensions, working temporary jobs well into old age in order to survive—these people have to get by in confined urban apartments with limited opportunity to engage beyond the four walls of their austere rooms. Older people like Ana-Maria (77), for instance, struggle to break up the monotony of life in the city. She lives alone in the huge flat that she bought with her husband in the late 1970s, upon returning from years of being allocated to work in the countryside. She lost her son when he was only 33 years old. Her daughter, four grandchildren and one great-grandchild all live in Australia. And although they support her financially, Ana-Maria and others in similar positions find fulfilment in day-trips with fellow pensioners. They take the bus to a hike in the local mountains surrounding the city. The highlight of the year tends to be a bigger trip—a few days at the beach, mostly, or, exceptionally, a couple of weeks abroad. Seasons do not make a difference to the everyday lives of older people bound to homes in the city—although, as would be expected, they note that they spend more time indoors during the autumn and winter months.

Similarly, those in the Village who cannot draw on the resources or competences needed for a translocal lifestyle are bound to making do with conditions of extreme weather, social and spatial isolation and very limited access to life-sustaining infrastructure. Over the years, Tsvetanka, for example, developed strict routines to help her cope with the hard winter months. She stores supplies during the warmer months (drawing on tested competences and skills of preserving, pickling and canning fruits, vegetables and meats; see Fig. 4.4) but still has to rely on the functioning of the bus for her monthly visit to the pharmacy in the nearby city:

**Fig. 4.4** A table set for afternoon tea during summer in the Village. It includes a choice of '*banitsa*' (filo pastry) and cake and the ubiquitous home-made jam. Milk and eggs are usually bought from local farmers or even the product of own cows or hens. During winter, the choice is often reduced to the bare minimum. (Photograph: Iossifova, July 2016)

> [In the winter,] I don't go out. Sometimes we've got two meters or so of snow, you can't go out. So I always have flour and yeast at home, in preparation, so that I can make bread. But even in the winter we have to go to the city to pick up medication and all that. (Tsvetanka, 85, July 2016)

Whether in the city or in the Village, therefore, for this group among Bulgaria's elderly translocal ageing is out of reach, whilst 'ageing in place' is but a trap. For a great many others, however, the four seasons differ greatly in terms of location and activities. Where people have villas in the countryside or by the sea, they divide their time between the city and the village, town or other city. On the one hand, it is more convenient to spend the winters in the city, where many have—at least in theory—access to central heating, whether or not they are able to afford it. This spares them the trouble of having to buy, prepare, store and carry firewood for heating and cooking in the countryside. On the other hand, houses in the countryside are often a lot more spacious than apartments in the city, which, in addition, many may have to share with adult children, grandchildren or others. Therefore, for some, the countryside offers a welcome escape from the environmental pressures of the city and the opportunity to engage with nature and live a healthier life:

I have a small house in [the countryside], where I go in the autumn and spring. It's too hot in the summer. I don't do anything there, I relax. (Svetla, 76, July 2016)

I grow tomatoes and I make pickles for the winter every time I go to [the countryside]. (Tanya, 76, June 2016)

In Sofia we feel more suppressed. I am not the type of person to go to the parks, coffee shops, restaurants and all that. So I spend a lot of time in the apartment. In the countryside, on the other hand, I spend my time outside, in a very good climate, so maybe that shows effect on our very good health. (Lyudmila, 86, July 2016)

As they age, older people encounter and adapt to situations where different factors prevent them from continuing the accustomed urban-rural oscillating on a weekly or seasonal basis. For instance, their own ill health or, as in the case of Raina, caring responsibilities may come in the way of frequent visits to the countryside:

We have a villa […], but since my husband got sick I don't go. Sometimes I go to clean up and make sure everything is alright, but just for a day. When he was still on his feet we used to go for two-three months every summer. We used to go with my grandson. The air is clean, it's beautiful. These days, there is not much difference between summer and winter. (Raina, 68, September 2016)

For those on tight pensions, budget considerations often lead them to abandon villas in the countryside. There, they argue, most everyday necessities are much more expensive, so staying in the city year-in, year-out is more affordable:

I stay here [in the city] because it's cheaper for me. If I go to the village—and I could go to the village, I have a place—the food is more expensive

[…]. I cannot afford it, so I stay here. […] Here, I have options, I can choose what's cheapest. Although, of course, the air is bad, the summers are hot, the winters are cold. (Yovka, 68, July 2016)

Similarly, the lack of options with regard to travelling and transportation are factors in the decision-making process. Where older people have had to sell their car or are no longer in a position to drive, reaching their homes in the countryside becomes a challenge and restricts how often they can go. A sense of loss and nostalgia come through when they speak about the times spent there in the past:

We had a car. But with the crisis—we had to sell it. It still pains me. But we had to sell it. I used to go to [the countryside] all the time. Now I have to rely on a neighbour who goes there twice a week, so he sometimes takes me. Otherwise, I wouldn't be able to go at all. (Tanya, 76, June 2016)

On the flip side, older people who do frequently oscillate between the city and the village mention that they find it difficult to initiate or even participate in activities they may be interested it, such as a choir or theatre group. When they join a regular activity, they have to abandon it, even if temporarily, as they travel to stay at their respective other location. In such cases, multi-local ageing is experienced as restrictive and counterproductive to processes of self-realisation. Similarly, physical and emotional labour is involved in compensating for being in one place rather than the other. An example of physical labour is the growing of seedlings on the balcony in the city for later planting in the countryside:

See, I take care of the plants. I have two balconies and keep plants there. In spring I start growing them [so that I can plant them in the garden when I go to] the villa. (Svetlana, 80, June 2016)

Equally, living translocally requires emotional labour, such as coping with the worries and fears of potential robberies and raids during times when older people are absent from their countryside homes:

I don't go [to the villa] in the winter, because I'm a bit afraid. Three-four times they robbed the house, so in the winter I'm a bit afraid because neighbours are not around to help [should something happen]. (Gergana, 74, July 2019)

Some older people in the Village, whilst not moving back and forth themselves, are part of multi-local families. They stay in the Village year-round, but share their homes with adult children during certain times of the year:

> My daughter lives here in the summer. She works here in the summer. In the winter I'm here alone. There are two flats in the house, it's a small house. In the winter, I am locked up in my room. (Nedelya, 69, July 2016)

What the previous paragraphs go to show is that a great many older people in Bulgaria are very familiar with the experience of oscillating between places, feeling equally attached to multiple places and, ultimately, ageing in multiple locations. Multi-local ageing is not a new phenomenon, but rather entrenched in older people's experiences as a result of Bulgaria's rural-urban reciprocity. Ever since the big shift to urbanisation in the 1950s, moving between the city and the village, creating multiple homes and ageing across multiple locations has been a key aspect of ageing in Bulgaria, and one that raises a series of questions with regard to the wide acceptance and use of the 'ageing in place' paradigm.

## BECOMING TRANSNATIONAL

In addition to the very common experience of multi-local ageing between the city and the village, the notions and experiences of sense of place and place attachment in Bulgaria are today complicated by the sharp rise of international migration. Nowadays, as is likely obvious by now from the narratives of older people throughout this book, almost everyone has family abroad. This includes adult children and grandchildren working or attending university abroad. Bogdan's circumstance is common among the new generations of older people:

> I have two children. [...] I have four grandchildren and two great-grandchildren. I hardly see them. One lives in Spain, another in London, another works in Greece, the fourth is here but just had a baby, so he's busy. (Bogdan, 94, July 2016)

Older people continue to expect the support of adult children and extended family abroad. This includes material, practical and emotional support. Material support denotes money as well as goods such as clothes and food. Practical support includes provision of sporadic care and small services, such as gardening, preparation of wood and so on. The latter can be—and very often is—provided by neighbours and friends, and the emotional support normally provided by close family can sometimes be replaced by meeting and talking with others, as discussed in Chap. 3.

> Of course, I rely most on my son and daughter here. But I have a niece in the States, my brother's daughter, in San Diego—she sends money when she can. […] They help me occasionally. (Albena, 86, July 2016)

In return, older people are prepared to help their adult children and grandchildren in whatever ways they can, even if it means increasing the already extraordinarily high levels of uncertainty in their own lives. This goes beyond simply offering practical advice or emotional support; it very often includes substantial sacrifice, such as giving up the safety and security of owning a home and not having to worry about paying the rent in the future:

> My sons were in the States for 15 years. They came back to open their own business. I sold my apartment […] and gave the money to them to help them start their business, and in exchange they promised to pay my rent. So I am renting. I live alone here. (Svetla, 76, July 2016)

Most commonly, adult children abroad enlist older people in looking after grandchildren, just as they would have had they stayed in Bulgaria. In the literature, the mobile parents (or even grandparents) of adult migrants are aptly referred to as G0 (*Zero generation*) (Wyss and Nedelcu 2018). This type of 'G0 parents' contribute to the provision of care and support to their children and grandchildren in the host country. They travel regularly back and forth and join the (migrant) household for shorter or longer periods of time, depending on childcare needs. They participate in the transmission of culture and family values and contribute substantially to transnational socialisation (Nedelcu 2012). However, this is not always or necessarily a positive experience for older people. For university-educated Nikolina, for instance, who has been living in the Village for 50-odd years, the memory of the one time she went to visit her son and his family abroad, in Belgium, is still haunting. She stayed two months and found it unbearable, despite the large number of Bulgarians there:

My days in Belgium, when I lived there for two months, were filled with reading, watching TV, cooking, waiting for the grandson to come home in the afternoon … It was really boring. I used to look out of the window for a bit of sunlight. But it was always dark and cloudy. So exactly two months after I arrived I was on my way back. Home! I don't want to go there ever again. Also, I am afraid of flying, and it's a real challenge for me. I took the bus. Two days and a night. I am too old for this. (Nikolina, 72, July 2016)

For older Bulgarians, as comes through in Nikolina's account, adjusting to a very different and oftentimes challenging climate in Western Europe (in comparison to the four seasons they are used to) presents an often overlooked trial. Temperature, relative humidity and cloudiness can show impact on ageing bodies and intensify the perception of pain, particularly among those suffering from rheumatism (Croitoru et al. 2019). This is in addition to the proven relationship between subjective wellbeing and the weather (Feddersen et al. 2012). Beyond the identity labour necessary to make sense of changes in one's everyday life and environment at the local level—as discussed in the previous chapters and sections—international migration adds additional burdens, such as the managing of different languages, beliefs, norms, administrative systems and other aspects entangled with everyday life (Božić 2006; Patterson 2004). Rositsa's account of G0 grandparenting demonstrates how these elements are compounded into an experience of the everyday that is characterised by isolation and fear, particularly where younger generations fail to respond to the emotional needs of the elderly:

In 2010 I went to England to help looking after my great-grandchildren. I was 75 then. […] I arrived there with one suitcase—I had only packed for a month. Overall, I stayed a year and a half. See, the grandchildren went to work, and I stayed with the baby. I was at home all day. Sometimes I took the risk to go outside with the baby. Thank god they don't know where I've taken him. You know, it's because the other grandmother looked after the baby, too, and one day she went out and went to a nearby park. So somebody called the police that there is a grandmother with a baby every day … So the police came and she had no documents with her, so they had to call the son … It's very different in England. Very isolated. I was going to die from nostalgia for the Village. I wanted to come home 1–2 months earlier than they had planned, but they told me no tickets were available. So I had to stay these last two months. It was awful. Now they wanted me to come back … They were thinking of having another, a third child! But I'm not going anymore. If I do, maximum 1–2 months. (Rositsa, 82, July 2016)

G0 grandparents usually visit their migrant children and grandchildren (predominantly adhering to a mother-daughter configuration) according to the following patterns: to offer family support when a baby is born, to provide childcare in planned and unplanned (emergency) situations, to provide permanent childcare support and to engage in 'intergenerational sharing and transmission, where the main purpose of the G0 visits is *Doing, enjoying and being together*' (Wyss and Nedelcu 2018). Stoyanka, whom we encountered in Chap. 2, spent the last eight years travelling back and forth between the Village and her granddaughter's home in Belgium—spending her winters abroad with family. Her experience clearly falls within the category of *doing, enjoying and being together*:

My granddaughter graduated from a Belgium university. [She has] a house there. When I'm in there, I have my own room. A typical day there would start at 8am, the children would already be at school. They drive them to school. I don't have any obligations in this regard. Sometimes I cook. At 5pm they pick them up from school. My grand-grandchildren speak French and English and Bulgarian. I try and speak Bulgarian to them all the time. I know a little bit of French, so I get by. It's enough to get by, though. Plus, there are a lot of Bulgarians, and even people from the Village.

When I'm here in the Village, after lunch I lie down a little to watch some movie or the other. Then I sit here on the bench, this friend or the other comes by for a chat. I've got friends here. Sometimes I go out to meet friends or to go to some event. But there are not many events now. I never go to bed before midnight. When I'm in Belgium, it's nice, too. They've got a good social life going with the mother-in-law there and friends, they come and visit all the time … But it's different in Belgium. One day we ran out of flour, so I said to my son-in-law: listen, I'm gonna go knock on the neighbour's door, they'll help me out. He was outraged—how? No! You can't just go and knock on people's doors … It's different, how people relate to each other there. Bulgarian customs don't work there. (Stoyanka, 78, July 2016)

In the context of transnational ageing, an analysis of the everyday reveals aspects of identity, sense of belonging as well as wider—structural—inequalities that can and should be addressed through policy interventions. Here, everyday life is not just about the 'mundane', but also about 'where we confront the wider structural questions of inequalities related to socio-economic differences, negotiate access to welfare and care institutions and deal with family and inter-generational relations. It is then as much about meso- and macro-level social structures, as it is about the experiences, questions of identity, emotions we attach and the meanings we give to our actions' (Näre et al. 2017).

Finally, I note the not unimportant role of digital technology in supporting older people to transcend notions of place altogether—albeit fleetingly (see Peine and Neven 2020). To stay in touch with family and friends, older people use mobile phones—usually received as a gift from younger generations. Landlines are ubiquitous but hardly ever used due to the convenience offered by more recent forms of communication. In the city, older people use computers and the internet primarily for the purpose of connecting with family abroad. Some very few possess the advanced skills that would be necessary to play computer games, for instance, or stay in touch with friends and former colleagues using social media. In the village, the picture is a bit trickier: computers and laptops are not frequently found, and where people have them in the house, they usually present relics left behind by grandchildren who have long moved out. Many appreciate that the use of new technology allows them to learn:

> For instance, [my son] said today—take a picture of the document and send it to me. Well, I can take a picture. But I don't know how to send it. So that's my new job. Every day, learning a little more about how to use [technology]. (Boyana, 67, June 2016)

Nedelcu (2017) accounts for the practices of grandparenting in the digital age and suggests that 'ICT-mediated family practices give rhythm and sense to the everyday life of elderly members of transnational families', but cautions that the emotional effects of such practices can be both gratifying and taxing. The latter is the case when hands-on care is needed but cannot be provided precisely because of physical distance; or when caring can otherwise be performed only through face-to-face contacts (Mason 2004).

> I talk to the children all the time, on the mobile, on the phone. We're in constant contact, skype, all that. I don't use the computer much—I'm out most of the day in the summer, and in the winter I don't go upstairs where

the computer is. Days in the wintertime are lonelier. (Nikolina, 72, July 2016)

However, although transnational families can be co-present across a range of settings, including the physical, virtual, by proxy and imagined (Baldassar 2007, 2008), hands-on care, such as the provision of childcare, is, of course, possible only when physically co-present. This insight contributes to feelings of sadness and regret among older people whose interactions with adult children and grandchildren abroad are limited to the virtual realm.

## CONCLUSIONS

In this chapter, I have argued that the notion of and agenda attached to 'ageing in place' is inherently unsuitable to capture and address the main concerns of older people in Bulgaria. Alluding to notions of mobility, I first show how people navigate familiar environments and then move to demonstrate that these environments are perpetually changing. In Bulgaria as in other rapidly transitioning societies uncertainty is normalised within the status quo. I then show that Bulgaria's rural-urban reciprocity has fostered profoundly translocal ageing processes, whereby older people oscillate between homes, social networks and environments in the city and in the village, thereby ageing in multiple and often radically different places. The ability to 'escape' the urban environment and spend time in the countryside is experienced as a beneficial factor and speaks to Bulgaria's typical rural-urban reciprocity. Finally, I show that recent trends in migration and processes of ageing are enmeshed, producing a kind of transnational ageing utterly incompatible with 'ageing in place' agendas.

Older people in Bulgaria today belong to a first generation to witness the mass exodus of younger generations in search of better lives abroad. They are the first generation to set up lives across national borders and even continents—beyond the limits of the former Soviet Bloc. They do so, I argue, to cater to the roles and expectations they have set for themselves within ever-shifting socioeconomic and cultural frames. As adult children and grandchildren emigrate to build livelihoods elsewhere, older people are prepared to do what their parents and grandparents have done—advise, support and provide care, but for them, to do so means to leave behind once more their homes, friends and familiar environments and to do so repeatedly and over prolonged, sometimes uncertain, periods of time. To do otherwise would mean to negate the value, norm and belief systems that they have built over the course of their lives—lives shaped by many dramatic disruptions, transitions and continued uncertainty. The fall of state socialism, and with it the accountability of the state, has meant a swift refocusing on family, friends and social networks as providers of support in times of need.

The willingness and ability of older people in Bulgaria to age translocally can therefore be interpreted as a mechanism for coping with the reality of not only being abandoned by a corrupt state, but also being left behind by adult children and grandchildren. Translocal ageing is a condition and a coping mechanism alike.

## REFERENCES

Atchley, R. C. (1971). Retirement and leisure participation: Continuity or crisis? *The Gerontologist, 11*(1_Part_1), 13–17.

Baker, P. M., & Prince, M. J. (1991). Supportive housing preferences among the elderly. *Journal of Housing for the Elderly, 7*(1), 5–24.

Baldassar, L. (2007). Transnational families and aged care: The mobility of care and the migrancy of ageing. *Journal of Ethnic and Migration Studies, 33*(2), 275–297.

Baldassar, L. (2008). Missing kin and longing to be together: Emotions and the construction of co-presence in transnational relationships. *Journal of Intercultural Studies, 29*(3), 247–266.

Benson, M. (2011). The movement beyond (lifestyle) migration: Mobile practices and the constitution of a better way of life. *Mobilities, 6*(2), 221–235. https://doi.org/10.1080/17450101.2011.552901.

Black, K. (2008). Health and aging-in-place: Implications for community practice. *Journal of Community Practice, 16*(1), 79–95. https://doi.org/10.1080/10705420801978013.

Borell, K. (2003). Family and household. *Family research and multi-household families. international review of sociology, 13*(3), 467–480.

Borglin, G., Edberg, A.-K., & Hallberg, I. R. (2005). The experience of quality of life among older people. *Journal of Aging Studies, 19*(2), 201–220.

Botterill, K. (2017). Discordant lifestyle Mobilities in East Asia: Privilege and Precarity of British retirement in Thailand. *Population, Space and Place, 23*(5), e2011. https://doi.org/10.1002/psp.2011.

Boyle, G. (2005). The role of autonomy in explaining mental ill-health and depression among older people in long-term care settings. *Ageing & Society, 25*(5), 731–748.

Boyle, A., Wiles, J. L., & Kearns, R. A. (2015). Rethinking ageing in place: The 'people' and 'place' nexus. *Progress in Geography, 34*(12), 1495–1511.

Božić, S. (2006). The achievement and potential of international retirement migration research: The need for disciplinary exchange. *Journal of Ethnic and Migration Studies, 32*(8), 1415–1427. https://doi.org/10.1080/13691830600928805.

Brenner, N. (2004). *New state spaces: Urban governance and the rescaling of statehood.* Oxford: Oxford University Press, Incorporated.

Carmona, M. (2010). *Public places, urban spaces: The dimensions of urban design.* London: Routledge.

Conkova, N., Vullnetari, J., King, R., & Fokkema, T. (2019). "Left like stones in the middle of the road": Narratives of aging alone and coping strategies in rural Albania and Bulgaria. *The Journals of Gerontology: Series B, 74*(8), 1492–1500.

Cristoforetti, A., Gennai, F., & Rodeschini, G. (2011). Home sweet home: The emotional construction of places. *Journal of Aging Studies, 25*(3), 225–232. https://doi.org/10.1016/j.jaging.2011.03.006.

Croitoru, A.-E., Dogaru, G., Man, T. C., Mălăescu, S., Motricală, M., & Scripcă, A.-S. (2019). Perceived influence of weather conditions on rheumatic pain in Romania. *Advances in Meteorology, 2019*, 9187105. https://doi.org/10.1155/2019/9187105.

Cutchin, M. P., Dickie, V., & Humphry, R. (2006). Transaction versus interpretation, or transaction and interpretation? A response to Michael barber. *Journal of Occupational Science, 13*(1), 97–99. https://doi.org/10.1080/1442759 1.2006.9686575.

Darling, J. (2009). Thinking beyond place: The responsibilities of a relational spatial politics. *Geography Compass, 3*(5), 1938–1954.

Deneva, N. (2012). Transnational aging Carers: On transformation of kinship and citizenship in the context of migration among Bulgarian Muslims in Spain. *Social Politics: International Studies in Gender, State & Society, 19*(1), 105–128. https://doi.org/10.1093/sp/jxr027.

Estes, C., Biggs, S., & Phillipson, C. (2003). *Social theory, social policy and ageing: A critical introduction.* Maidenhead: Open University Press.

Europe's first 'smart city' to land in Bulgaria. (2018, July 6). *CGTN.*

European Commission. (2018). *Active ageing.* Retrieved January 2, 2018, from http://ec.europa.eu/social/main.jsp?catId=1062&langId=en.

Feddersen, J., Metcalfe, R., & Wooden, M. (2012). *Subjective well-being: Weather matters; climate Doesn't.* Melbourne: Melbourne Institute.

Francis, M. (1989). Control as a dimension of public-space quality. In I. Altman & E. H. Zube (Eds.), *Public places and spaces* (pp. 147–172). Boston, MA: Springer US.

Fried, M. (2000). Continuities and discontinuities of place. *Journal of Environmental Psychology, 20*(3), 193–205.

Gesler, W. M., & Kearns, R. A. (2005). *Culture/place/health* (Vol. 16). London and New York: Routledge.

Gory, M. L., Ward, R., & Sherman, S. (1985). The ecology of aging: Neighborhood satisfaction in an older population. *The Sociological Quarterly, 26*(3), 405–418.

Hillcoat-Nallétamby, S. (2014). The meaning of "Independence" for older people in different residential settings. *The Journals of Gerontology: Series B, 69*(3), 419–430. https://doi.org/10.1093/geronb/gbu008.

Horn, V., & Schweppe, C. (2017). Transnational aging: Toward a transnational perspective in old age research. *European Journal of Ageing, 14*(4), 335–339. https://doi.org/10.1007/s10433-017-0446-z.

Horn, V., Schweppe, C., & Um, S.-G. (2013). Transnational aging—A young field of research. *Transnational Social Review, 3*(1), 7–10. https://doi.org/1 0.1080/21931674.2013.10820744.

Howden-Chapman, P., Signal, L., & Crane, J. (1999). Housing and health in older people: ageing in place. *Social Policy Journal of New Zealand* (13), 14–30.

Johansson, K., Laliberte Rudman, D., Mondaca, M., Park, M., Luborsky, M., Josephsson, S., et al. (2013). Moving beyond 'aging in place' to understand migration and aging: Place making and the centrality of occupation. *Journal of Occupational Science, 20*(2), 108–119. https://doi.org/10.1080/1442759 1.2012.735613.

King, R., & Vullnetari, J. (2006). Orphan pensioners and migrating grandparents: The impact of mass migration on older people in rural Albania. *Ageing and Society, 26*(5), 783–816. https://doi.org/10.1017/S0144686X06005125.

Konstantinov, Y., & Simić, A. (2001). Bulgaria: The quest for security. *Anthropology of East Europe Review, 19*(2), 21–34.

Lawton, M. P. (1985). The elderly in context: Perspectives from environmental psychology and gerontology. *Environment and Behavior, 17*(4), 501–519.

Machlis, G. E., Force, J. E., & Burch, W. R. (1997). The human ecosystem part I: The human ecosystem as an organizing concept in ecosystem management. *Society & Natural Resources: An International Journal, 10*(4), 347–367.

Martin, G. P., Nancarrow, S. A., Parker, H., Phelps, K., & Regen, E. L. (2005). Place, policy and practitioners: On rehabilitation, independence and the therapeutic landscape in the changing geography of care provision to older people in the UK. *Social Science & Medicine, 61*(9), 1893–1904.

Mason, J. (2004). Managing kinship over long distances: The significance of 'the visit'. *Social Policy and Society, 3*(4), 421–429. https://doi.org/10.1017/S1474746404002052.

Means, R., & Evans, S. (2012). Communities of place and communities of interest? An exploration of their changing role in later life. *Ageing and Society, 32*(8), 1300–1318. https://doi.org/10.1017/S0144686X11000961.

Milenkovic, A. (2018, 6 July 2018). Europe's first 'smart city' to land in Bulgaria. CGTN. Retrieved from https://news.cgtn.com/news/3d3d674e346354 4e78457a6333566d54/share_p.html.

Milligan, C. (2012). *There's no place like home: Place and care in an ageing society.* Farnham: Ashgate Publishing, Ltd.

Müller, M. (2020). In search of the global east: Thinking between north and south. *Geopolitics, 25*(3), 734–755. https://doi.org/10.1080/1465004 5.2018.1477757.

Näre, L., Walsh, K., & Baldassar, L. (2017). Ageing in transnational contexts: Transforming everyday practices and identities in later life. *Identities, 24*(5), 515–523. https://doi.org/10.1080/1070289X.2017.1346986.

Nedelcu, M. (2012). Migrants' new transnational habitus: Rethinking migration through a cosmopolitan lens in the digital age. *Journal of Ethnic and Migration Studies, 38*(9), 1339–1356.

Nedelcu, M. (2017). Transnational grandparenting in the digital age: Mediated co-presence and childcare in the case of Romanian migrants in Switzerland and Canada. *European Journal of Ageing, 14*(4), 375–383. https://doi.org/10.1007/s10433-017-0436-1.

Nikolov, N. (2020). *The panelka palimpsest: Transformation of everyday life in a prefabricated neighbourhood in Sofia.* PhD dissertation, University College London, London.

Oldman, C., & Quilgars, D. (1999). The last resort? Revisiting ideas about older people's living arrangements. *Ageing & Society, 19*(3), 363–384.

Parusheva, D., & Marcheva, I. (2010). Housing in socialist Bulgaria: Appropriating tradition. *Home Cultures, 7*(2), 197–215. https://doi.org/10.275 2/175174210X12663437526214.

Patterson, F. M. (2004). Policy and practice implications from the lives of aging international migrant women. *International Social Work, 47*(1), 25–37. https://doi.org/10.1177/0020872804039368.

Paulsen, K. E. (2010). Placemaking. In R. Hutchison (Ed.), *Encyclopedia of urban studies* (pp. 600–603). Thousand Oaks, CA: SAGE Publications.

Peace, S., Holland, C., & Kellaher, L. (2011). 'Option recognition' in later life: Variations in ageing in place. *Ageing and Society, 31*(5), 734–757. https://doi.org/10.1017/S0144686X10001157.

Peine, A., & Neven, L. (2020). The co-constitution of ageing and technology—A model and agenda. *Ageing and Society,* 1–22. https://doi.org/10.1017/S0144686X20000641.

Phillipson, C. (2004). Urbanisation and ageing: Towards a new environmental gerontology. *Ageing & Society, 24*(06), 963–972. https://doi.org/10.1017/S0144686X04002405.

Phillipson, C., & Ahmed, N. (2006). Transnational communities, migration and changing identities in later life: A new research agenda. In S. O. Daatland & S. Biggs (Eds.), *Ageing and diversity. Multiple pathways and cultural migrations* (pp. 157–172). Bristol: Policy Press.

Rowles, G. D. (1993). Evolving images of place in aging and 'aging in place'. *Generations: Journal of the American Society on Aging, 17*(2), 65–70.

Schier, M. (2016). Everyday practices of living in multiple places and Mobilities: Transnational, Transregional, and intra-communal multi-local families. In M. Kilkey & E. Palenga-Möllenbeck (Eds.), *Family life in an age of migration and mobility: Global perspectives through the life course* (pp. 43–69). London: Palgrave Macmillan UK.

Schwanen, T., Hardill, I., & Lucas, S. (2012). Spatialities of ageing: The co-construction and co-evolution of old age and space. *Geoforum, 43*(6), 1291–1295. https://doi.org/10.1016/j.geoforum.2012.07.002.

Severinsen, C., Breheny, M., & Stephens, C. (2016). Ageing in unsuitable places. *Housing Studies, 31*(6), 714–728. https://doi.org/10.1080/0267303 7.2015.1122175.

Smith, A. E. (2009). *Ageing in urban neighbourhoods: Place attachment and social exclusion*. Bristol: Policy Press.

Smollett, E. W. (1985). Settlement systems in Bulgaria: Socialist planning for the integration of rural and urban life. In A. Southall, P. J. M. Nas, & G. Ansari (Eds.), *City and society: Studies in urban ethnicity, life-style and class*. Leiden: Institute of Cultural and Social Studies, University of Leiden.

Staddon, C., & Mollov, B. (2000). City profile: Sofia, Bulgaria. *Cities, 17*(5), 379–387. https://doi.org/10.1016/S0264-2751(00)00037-8.

Tomorrow's smart city begins in South European country. (2017, May 15). *China Daily*.

Wiles, J. (2005). Conceptualizing place in the care of older people: The contributions of geographical gerontology. *Journal of Clinical Nursing, 14*, 100–108.

Wyss, M., & Nedelcu, M. (2018). Zero generation grandparents caring for their grandchildren in Switzerland. The diversity of transnational care arrangements among EU and non-EU migrant families. In V. Ducu, M. Nedelcu, & A. Telegdi-Csetri (Eds.), *Childhood and parenting in transnational settings* (pp. 175–190). Cham: Springer International Publishing.

Xu, H., & Wu, Y. (2016). Lifestyle mobility in China: Context, perspective and prospects. *Mobilities, 11*(4), 509–520. https://doi.org/10.1080/1745010 1.2016.1221027.

Zhou, Y. R. (2012). Space, time, and self: Rethinking aging in the contexts of immigration and transnationalism. *Journal of Aging Studies, 26*(3), 232–242. https://doi.org/10.1016/j.jaging.2012.02.002.

Zhuan, T. (2017, 15 May 2017). Tomorrow's smart city begins in South European country. China Daily. Retrieved from http://www.chinadaily.com.cn/cndy/2017-05/15/content_29345007.htm.

Zimmer, Z., Rada, C., & Stoica, C. A. (2014). Migration, location and provision of support to older parents: The case of Romania. *Journal of Population Ageing, 7*(3), 161–184.

# CHAPTER 5

# Trajectories of Ageing: Learning (from) the Global East

**Abstract** In this chapter, Iossifova offers a brief overview of the empirical findings and main arguments presented in *Translocal Ageing in the Global East*. She cautions against the 'ageing in place' paradigm, arguing that for older people in Bulgaria and elsewhere in the Global East, it is profoundly out of step with their translocal experience of ageing. She calls for more and deeper engagement with the Global East and its intersecting experiences of post-socialism, rapid urbanisation, rural-to-urban and outmigration as well as socio-spatial restructuring. Finally, she reflects on the use of the human ecosystem framework to structure comprehensive research on human-environment interactions in ageing studies and the social sciences.

**Keywords** Bulgaria • Global East • 'Ageing in place' • Translocal ageing • Resilience • Older people • Human ecosystem framework • Grounded theory

When I first arrived in Shanghai in the spring of 2004, I was overwhelmed. Everything seemed very different. The endless swaths of high-rise buildings that we passed as the driver took me from the airport to the city; the odours of unaccustomed foods and fragrant plants; the whitish haze of the early morning hours, and, preceding it, the darkness of the dimly lit streets; the incessant and unfamiliar sounds of life in a metropolis that never seemed to sleep. Every day, as I traversed the city from my apartment to

© The Author(s) 2020
D. Iossifova, *Translocal Ageing in the Global East*,
https://doi.org/10.1007/978-3-030-60823-1_5

my place of work, I watched uniformed students exercise to patriotic songs before school start; I participated in the community-building activities regularly organised by my office; I listened to the curious chatter and patter that filled my street just before dinner time; I peeked through windows and doors that seemed always ajar; and I stopped to attempt chats with the older women and men who sat on small chairs outside my building, some wearing red armbands to ensure they are recognised as members of the neighbourhood committee. And then, one day, it hit me. Having spent the previous two decades of my life away from Bulgaria, here, in Shanghai, I felt at home again.

\*    \*    \*

In this brief final chapter, I have two main aims. The first is, as would be expected for the concluding section of a book, to summarise its key empirical, methodological and theoretical contributions. The second is, as the opening paragraph suggests, to locate the material presented here within the wider context of the Global East and to argue for (fresh) engagement with the geographical and conceptual domain. The Global East holds unique lessons for scholarship and practice at the interface of ageing, transitions, public policy and urbanism, among a myriad of other relevant fields.

I begin by giving a brief overview of the empirical findings and main arguments presented in the preceding chapters; I then caution against the 'ageing in place' paradigm and argue for more and deeper engagement with the Global East and its intersecting experiences of rapid urbanisation, rural-to-urban and outmigration, socio-spatial restructuring and postsocialism in its many guises. In closing, I turn to revisit and reflect on the project's methodological approach and its usefulness in applied social science research and ageing studies, in particular.

## Lifecourse, Everyday Life and Translocal Geographies of Ageing Under Uncertainty

Older people in Bulgaria's cities and villages today are surprisingly resilient in the face of mounting financial, economic, ecological, social and cultural challenges. I define resilience here in the ecological sense of the term as the capacity to face adversity, remain well and even thrive (Ryff et al. 1998). I attribute this unexpected resilience to three key factors: (1) the experience of Bulgaria's older people today of a series of dramatic transformations straddling all domains of human live; (2) the recent decades of Transition towards market economy and a semblance of democracy and the now accustomed state of permanent uncertainty; and (3) the normalisation of translocal ageing linked to Bulgaria's rural-urban reciprocity (Smollett 1985; Konstantinov and Simić 2001) and associated processes of life-long oscillating between the city and the countryside.

Today's older people have lived through two historical periods associated with major socioeconomic, spatial and political transitions. As I demonstrated in Chap. 2, the lives of today's older people were marked by the political, economic, social and intimate circumstances of the Communist era (1944–1989) and the long-lasting Transition to democracy and market as organising systems following the fall of state socialism. These transitions had a profound impact on the lifecourse, identities, roles and everyday lives of older people. They were experienced as major disruptions to accustomed value, norm and belief systems. Today's older people grew up weighing themselves in the knowledge that the state took care of everything and everybody, thus acknowledging the limits to their ability to influence the course of their own lives beyond decisions of marginal importance (although the state had gained control even of the most minor and insignificant of detail).

The recent decades of Transition towards market economy and a semblance of democracy have brought about the now accustomed state of permanent uncertainty which, in turn, demanded initiative and agency from older people if they were to remain well and even thrive (Ryff et al. 1998). The fall of state socialism triggered a series of processes, such as restitution (displacing many from their long-term homes), unemployment or change in employment, and the radical withdrawal of the state from the provision of previously taken-for-granted public services, such as child care. Importantly, the Transition forced ageing people in Bulgaria to unlearn their reliance on the state and to develop, instead, the initiative and agency that are necessary to navigate a newly emerging and constantly shifting socioeconomic and political landscape. In Chap. 3 I argued that it is this kind of agency that makes it possible for today's older people to manage their everyday lives under savage capitalism and in a state of constant uncertainty. I worked through unequal infrastructures of provision and how they are linked with older people's patterns of everyday life, subsequently focusing on the strategies deployed to cope with meagre pensions, ailing bodies and tattered homes. I noted that older people today refocus relationships of trust to the more traditional assemblages of family and friends who they now turn to for support in times of need. Finally, I outlined the differences in patterns of everyday life between the city and the Village to argue that—whilst ageing in the Village certainly poses the greater challenges—older people in the city *and* the Village display enormous adaptive resilience in their everyday lives.

I argue that Bulgaria's rural-urban reciprocity (Smollett 1985; Konstantinov and Simić 2001) and associated processes of life-long oscillation between the city and the countryside have led to the normalisation of translocal ageing, thus paving the way for the more recent phenomenon of transnational ageing. In Chap. 4, I revisited the notion of 'ageing in place' and the concept of rural-urban reciprocity (Smollett 1985; Konstantinov and Simić 2001) and the processes of translocal ageing associated with such rural-urban entanglement. I presented briefly how older people navigate their familiar environments in the city and the Village before arguing that these may not be as familiar, after all, due to the continuous neoliberal restructuring of the social and built environments (see also Hirt 2012). Over the course of their lives, older people have constructed a mode of oscillating between the city and the countryside that has instilled versions of multi-local and translocal lifestyles. This has made it easier for them to adapt to the requirements and expectations emerging from the newly acquired international mobility of their children and grandchildren. For older people in Bulgaria today, transnational ageing— that is, oscillating not only between the city and the countryside but also between the city, the countryside and the rest of the world—is at the heart of their experience of ageing.

## Resolving the Dogma of 'Ageing in Place'

As should be quite obvious by this point, at the core of this book is my critique of the notion of 'ageing in place'. Today, it is widely believed that a positive ageing experience is the result of the physical (and economic) independence and identity continuity associated with ageing in one's lifetime neighbourhood and home, or 'ageing in place' (e.g., Wiles 2005; Peace et al. 2011). At first glance, the arguments behind 'ageing in place' appear benign. On closer inspection, however, the policy agendas that sprout from the notion of 'ageing in place' are all but benevolent.

'Ageing in place' underpins WHO policy guidelines and serves to guide planning and design, for instance, in the guise of 'age-friendly cities'

(Buffel et al. 2012; Buffel and Phillipson 2018). Adopting the 'ageing in place' framework uncritically for application around the globe is running the risk of falling prey to a Trojan horse. To begin with, the paradigm is not always the right fit. It neglects the complex translocal experiences of older people who, for whatever reasons, age away from their homes. It disregards those with histories of rural-to-urban or international migration and those who are part of multi-local or transnational families.

The territorial-bounded thinking that assumes 'a quasi-natural overlap between a territory and an individual's social space' (Horn and Schweppe 2017; Zhou 2012) is profoundly outdated. Rather, it is time to reflect on the complex ageing experiences of older people in the specific context(s) within which they age, conceptualising the human-environment relationship as dynamic and taking into account the importance of translocal ageing.

'Ageing in place' is aligned with the neoliberal principles upon which most capitalist societies in the Global North have built their socioeconomic systems. Consistent with the neoliberal project of responsibilisation as the 'new emphasis on the personal responsibilities of individuals, their families and their communities for their own future well-being and upon their own obligation to take active steps to secure this' (Rose 1996), the 'ageing in place' paradigm puts the individual in charge of their own health and wellbeing and encourages the retreat of the state from responsibility and accountability towards its ageing population.

Accepting that older people should age independently in place, without the support of the state or even family and friends, may mean accepting that they should be fully in charge of their livelihoods; that they should remain economically independent, earning and spending income well into old age. 'Ageing in place' is simply not an option for the majority of people outside wealthy former 'First World' countries; it is something that the ageing populations of the Global East (Müller 2020) or Global South can neither afford nor do they have the (questionable) luxury to desire.

## LEARNING (FROM) THE GLOBAL EAST

Whilst attempting to focus strictly on Bulgaria, choosing to refrain from intentional or in-depth comparison between different geographical contexts, in the process of writing this book I have found myself frequently drawing parallels between China and Bulgaria. In Chap. 2, for instance, I note similarities in movement and residency restrictions during socialism;

the design and construction of housing; basic elements of functioning in society (such as *guanxi/vraski*). The likenesses are striking. My experience leads me to believe that ageing in Bulgaria has likely much more in common with ageing in China or Vietnam than it has with ageing in Britain, for instance. This may well be attributed to the shared experience of communism and state socialism, rapid rural-to-urban migration and the more recent rise in international migration (e.g., Liu 2014). In addition, as I have hinted here and there throughout the book, parallels exist in cultural specificities such as the historical importance of the extended family (Cheung et al. 1980; Logan et al. 1998), the continuous presence of multi-generation households (Ran and Liu 2020) or the nature of inter-generational relationships (Jackson and Liu 2017).

Modern-day Bulgaria is willing to accept Chinese investment as part of the People's Republic's Belt and Road Initiative, interpreted by some as the attempt to 'expand Beijing's international influence and trade abroad in order to compensate for the slowdown in economic growth at home' (Cheresheva 2017). Projects such as Saint Sofia, a China-backed smart city on the outskirts of Bulgaria's capital (briefly mentioned in Chap. 4), suggest that China's influence in Bulgaria can only be expected to grow in the future (Vangeli 2017). Along the way, interventions of this kind will show palpable and long-lasting impact, such as transitions in urban form. The smart city Saint Sofia proposed to replace swaths of sunflower fields had very little in common with local architectural conventions; rather, plans represented the out-of-scale copy-paste features of China's high-speed urbanism, which are becoming characteristic of new-built settlements in the Global East (Cheresheva 2017; Milenkovic 2018).

All this provides much ground for speculation that 'the Global East(s)' (Müller 2020, 2021) is *en route* to become an umbrella for a range of contexts evolving along similar trajectories. A comparative lens can bring much-needed insight and improve our understanding of similarities, differences and possible future pathways across multiple domains of inquiry, including at the intersection of urbanism and ageing studies, as they relate to the Global East. Global East scholarship questions the North-South dichotomy, especially as calls for a much more cautious reading of the world (and its habitually marginalised geographies) can no longer be muffled (Barker and Pickerill 2020). Scholars recognise that there is need for differentiation within the vast space that remains to be claimed once the Global North has been delimited along the boundaries of Western Europe and North America.

The notion of the Global East remains vague and is defined in different ways. For instance, in their direction-setting work on gentrification, Shin et al. (2016) consider the Global East to comprise some countries in the geographically relatively narrow East and Southeast Asia—Hong Kong, Singapore, South Korea and Taiwan, the Philippines and Indonesia as well as China and Vietnam. They note that 'these countries display some share experiences such as rapid urbanisation, export-oriented economic development and strong developmental states with authoritarian pasts or inclinations' yet very different urbanisation outcomes due to 'place-based geographical and historical specificities' (Shin et al. 2016). I sympathise with Müller's (2021) broader and yet more precise definition of the 'Global Easts' as 'all those places that do not easily fit into the categories of Global North and Global South', including those in the Middle East, East Asia and the former Soviet Bloc. Their shared history is rooted in their experiences of being either a non-European imperial power (e.g., Russia and China) or a colony of non-European empires (such as Bulgaria and its 500 years under Ottoman occupation). These experiences position places of the Global East firmly outside the influence of European colonialism (Müller 2021).

Understanding the conditions and transformations of the Global East requires 'theorising, distorting, mutating and bringing into question' (Shin et al. 2016) the entrenched concepts, frameworks and theories of the Global North. Engaging with the Global East in scholarship and practice cannot mean 'simply "expanding" towards the Global East (periphery) through colonisation or one-way policy transfer, as if [engaging with] an imported new phenomenon' (Shin et al. 2016). Rather, the currently unfolding decolonial turn in human geography (and beyond) 'builds on and extends postcolonial, feminist and critical race geography by centring the forms of knowledge production under colonial-modernity, in order to refine understandings of its particularities and to reanimate critiques of racialisation, colonial-modern resource distributions and epistemic violence' (Radcliffe 2017). It 'encourages re-thinking the world *from* Latin America, *from* Africa, *from* Indigenous places and *from* the marginalised academia in the global South' (Radcliffe 2017). Of course, this should expand to encourage re-thinking the world *from* the Global Easts. I join in with Müller's (2021) call for 'an active engagement with the wealth of literature that is out there, whether on the postsocialist or on other Easts'.

## REFLECTIONS ON METHODOLOGY

Echoing above call for 'theorising, distorting, mutating and bringing into question' (Shin et al. 2016) established concepts, frameworks and theories of the Global North, in this brief section I reflect on the methodology underpinning my research, and, in particular, on the use of the human ecosystem framework (HEF) (Machlis et al. 1997) as the starting point and structuring device in my research process.

The research started with an interest in the coping mechanisms of older people in Bulgaria. In a typical grounded theory fashion (Glaser and Strauss 1967), I was keen to discover the main concerns of people on the ground whilst avoiding to be influenced, more than strictly necessary, by the extant literature or preconceived hypotheses or questions. Practical considerations, such as limitations to my funding or the requirement to have ethics approval in place before travelling to Bulgaria, led to the decision to devise an interview guide that would cover the broadest possible range of experiences and concerns without anticipating or constraining research participants' responses.

Choosing the human ecosystem (Machlis et al. 1997) as the underlying framework for the design allowed me to do precisely that. In hindsight, the framework proved ideal for the purpose. The resulting 25-question interview guide covered human (social) systems through the categories of social order, social cycles and social institutions and subjects aligned with HEF component parts (such as identity, health or physiological cycles); it covered critical resources through the categories of natural resources, cultural resources and socioeconomic resources and the subjects of energy, organisations and capital, among various others. Respondents were free to reply in as much or as little detail as they wished—and usually, the brief questions asked led to detailed responses and rich narratives that revealed the extent to which older people were entangled with the different component parts of the human ecosystem of which they themselves were part.

Interestingly, relatively 'dry' or technical questions often triggered unexpectedly emotional and long-winded responses. For instance, the simple question of 'How do you heat your home?' (to cover the component part of 'energy' under the 'natural resources' category) could be answered with a simple 'I have central heating' (which we then probed further); or it could produce narratives around 'heating careers' with respondents looking back to the different technologies and media they used over the course of their lives; when and where; and what kind of cognitive or emotional associations they had with the respective heating modes, indicating the multiple and intricate ways in which the component parts of human ecosystems are linked with people's individual and social histories, everyday lives and envisioned or desired futures.

In a nutshell, simply in that it provides structure to the component parts of human ecosystems—rather than theories about their inherent properties, let alone their evolving dynamics—the HEF can be used by researchers working in many kinds of traditions. The framework is marked by a much-needed ideological neutrality and therefore particularly well suited for application in studies subscribed to the use of decolonising methodologies, be they quantitative, qualitative or mixed (Smith 1999). That is, methodologies that include non-Western alongside Western thought in systems of knowledge, knowledge discovery and knowledge production (Mignolo 2010).

Of course, it could be argued here that the HEF stands in direct opposition to the grounded theory approach, or that—developed in the United States—it is a framework from and for the analysis of human ecosystems of the Global North, therefore contradicting the argument made above for a scholarship of and from the Global East. My use of the framework served the purpose of going through the register of human and non-human component parts of the human ecosystem and querying domains of the life-course and everyday life that I may not have, otherwise, considered. In this way, the HEF proved invaluable in designing a comprehensive interview guide for the collection of data.

Conceptualisation and theory-finding happened only during the subsequent constant comparative analysis of the collected data, and I should note that here, too, I followed a grounded theory approach in accepting that 'all is data' (Glaser 2002). This dictum gave me the freedom to reflect on and work through the narratives of interview participants; my own observations during visits to Bulgaria; my personal and intimate experiences and accounts; informal conversations with friends, relatives and acquaintances; and many, many other kinds of data that became part of and influenced the analysis presented here.

## In Closing

The motivation behind this book came from visiting Bulgaria and observing older people in (not so) public spaces: exhausted older people working shifts as shop assistants; dragging their visibly aching bodies onto public transport; watching the outside world from their windows; even begging in the streets, bowing as pedestrians pass by. Acknowledging the forces of massive emigration, leading the country to lose almost one quarter of its population over the course of 30 years, combined with very low pensions, continuously rising living costs and an unkempt, crumbling housing stock, my initial intention was to capture and document what I expected to be the lonely and desolate lives of a generation of abandoned misers.

In Bulgaria, as elsewhere in the Global East, the collective memory of the state that takes care of everything and everyone has not yet fully faded. Indeed, as I have shown throughout this book, older people in Bulgaria have been abandoned by a state that had once made that promise. Yet, in the face of unprecedented challenges in every domain of their lives, Bulgaria's older people display astounding levels of resilience. As I complete this book, the country is entering yet another state of crisis. After weeks of protests—and amid a raging COVID-19 epidemic—the country's long-serving Prime Minister, Boyko Borissov, promised to 'restart the state' (Dimitrov 2020). That is, rewriting the constitution, halving the number of members of parliament and reforming the judiciary. It remains to be seen what this will mean in concrete terms; it is already clear, however, that for Bulgaria's older people the state of perpetual uncertainty is a far cry from coming to an end.

Therefore, I tread carefully as I paint a picture of the triumph of resilience over uncertainty, hardship and abandonment. There is a thin line between older people's newly activated agency as 'the capacity that an individual acquires to plan his/her own future assuming an active, conscious and intentional role in achieving this future' (Romaioli and Contarello 2019)—and their deliberate responsibilisation in the absence of a responsible and accountable state. Although Bulgaria's older people have proved incredibly resilient, this should not be regarded 'as the possibility of transformative politics that exists in situations of heightened suffering' (Lesutis 2019). Hall (2019) brilliantly argues that 'investment in social infrastructure is one way of realising a politics of care in times of austerity. This investment is also interpersonal and relational, since it centres social and political responsibilities and institutions. However, the retraction of public expenditure and a lack of investment in social infrastructure do not necessarily result in care-less social relations'. Yet, the state needs to be held accountable.

Bulgaria's older people may well have been abandoned. However, they have developed and applied adaptive strategies that allow them to cope with the state of perpetual uncertainty characterising every domain of their lifecourse and everyday lives. Within the few thousand words of this book, I have presented how older people's lifecourse and patterns of everyday life are entangled with the historical, social, political and geographical contexts within which they occur. Their resilience is woven into the complex fabric of translocal and intergenerational relationships that I have sought to portray in this book. In many instances, I have just touched upon and most certainly not given due attention to interesting and important aspects of ageing in Bulgaria. It is therefore my heartfelt hope that this book provides encouragement and inspiration for future engagement in research and practice, with Bulgaria and with the wider Global Easts.

## REFERENCES

Barker, A. J., & Pickerill, J. (2020). Doings with the land and sea: Decolonising geographies, Indigeneity, and enacting place-agency. *Progress in Human Geography, 44*(4), 640–662. https://doi.org/10.1177/0309132519839863.

Buffel, T., & Phillipson, C. (2018). A manifesto for the age-friendly movement: Developing a new urban agenda. *Journal of Aging & Social Policy, 30*(2), 173–192. https://doi.org/10.1080/08959420.2018.1430414.

Buffel, T., Phillipson, C., & Scharf, T. (2012). Ageing in urban environments: Developing 'age-friendly' cities. *Critical Social Policy, 32*(4), 597–617. https://doi.org/10.1177/0261018311430457.

Cheresheva, M. (2017, 9 August 2017). Bulgaria Backs New Chinese 'Smart City' Near Sofia. BalkanInsight. Retrieved from http://www.balkaninsight.com/en/article/bulgaria-extends-support-for-chinese-smart-city-near-sofia-08-08-2017

Cheung, L. Y., Cho, E. R., Lum, D., Tang, T. Y., & Yau, H. B. (1980). The Chinese elderly and family structure: Implications for health care. *Public Health Reports, 95*(5), 491–495.

Dimitrov, M. (2020). How a bizarre beach plot landed Bulgaria's longtime PM in hot water. *The Guardian,*

Milenkovic, A. (2018, 6 July 2018). Europe's first 'smart city' to land in Bulgaria. CGTN. Retrieved from https://news.cgtn.com/news/3d3d674e346354 4e78457a6333566d54/share_p.html

Glaser, B. G. (2002). Constructivist grounded theory? *Forum Qualitative Sozialforschung/Forum: Qualitative Social Research, 3*(3), Art. 12.

Glaser, B. G., & Strauss, A. L. (1967). *The discovery of grounded theory: Strategies for qualitative research.* New York: Aldine Transaction.

Hall, S. M. (2019). *Everyday life in austerity: Family, friends and intimate relations.* Cham: Palgrave Macmillan.

Hirt, S. (2012). *Iron curtains: Gates, suburbs and privatization of space in the post-socialist city*. Chichester: John Wiley & Sons.

Horn, V., & Schweppe, C. (2017). Transnational aging: Toward a transnational perspective in old age research. *European Journal of Ageing, 14*(4), 335–339. https://doi.org/10.1007/s10433-017-0446-z.

Jackson, S., & Liu, J. (2017). The social context of ageing and intergenerational relationships in Chinese families. *The Journal of Chinese Sociology, 4*(1), 2. https://doi.org/10.1186/s40711-016-0050-1.

Konstantinov, Y., & Simić, A. (2001). Bulgaria: The quest for security. *Anthropology of East Europe Review, 19*(2), 21–34.

Lesutis, G. (2019). The non-politics of abandonment: Resource extractivisim, precarity and coping in Tete, Mozambique. *Political Geography, 72*, 43–51. https://doi.org/10.1016/j.polgeo.2019.03.007.

Liu, J. (2014). Ageing, migration and familial support in rural China. *Geoforum, 51*, 305–312. https://doi.org/10.1016/j.geoforum.2013.04.013.

Logan, J. R., Bian, F., & Bian, Y. (1998). Tradition and change in the urban Chinese family: The case of living arrangements. *Social Forces, 76*(3), 851–882. https://doi.org/10.1093/sf/76.3.851.

Machlis, G. E., Force, J. E., & Burch, W. R. (1997). The human ecosystem part I: The human ecosystem as an organizing concept in ecosystem management. *Society & Natural Resources: An International Journal, 10*(4), 347–367.

Mignolo, W. (2010). Cosmopolitanism and the De-colonial option. *Studies in Philosophy and Education, 29*(2), 111–127. https://doi.org/10.1007/s11217-009-9163-1.

Müller, M. (2020). In search of the global east: Thinking between north and south. *Geopolitics, 25*(3), 734–755. https://doi.org/10.1080/14650045.2018.1477757.

Müller, M. (2021). Footnote urbanism: The missing east in (not so) global urbanism. In M. Lancione & C. McFarlane (Eds.), *Thinking global urbanism: Essays on the City and its future*. London: Routledge.

Peace, S., Holland, C., & Kellaher, L. (2011). 'Option recognition' in later life: Variations in ageing in place. *Ageing and Society, 31*(5), 734–757. https://doi.org/10.1017/S0144686X10001157.

Radcliffe, S. A. (2017). Decolonising geographical knowledges. *Transactions of the Institute of British Geographers, 42*(3), 329–333. https://doi.org/10.1111/tran.12195.

Ran, G. J., & Liu, L. S. (2020). 'Forced' family separation and inter-generational dynamics: Multi-generational new Chinese immigrant families in New Zealand. *Kōtuitui: New Zealand Journal of Social Sciences Online*, 1–20. https://doi.org/10.1080/1177083X.2020.1801772.

Romaioli, D., & Contarello, A. (2019). Redefining agency in late life: The concept of 'disponibility'. *Ageing and Society, 39*(1), 194–216. https://doi.org/10.1017/S0144686X17000897.

Rose, N. (1996). The death of the social? Re-figuring the territory of government. *Economy and Society, 25*(3), 327–356. https://doi.org/10.1080/03085149600000018.

Ryff, C. D., Love, G. D., Essex, M. J., & Singer, B. (1998). Resilience in adulthood and later life. In J. Lomranz (Ed.), *Handbook of aging and mental health* (pp. 69–96). New York: Plenum Press.

Shin, H. B., Lees, L., & López-Morales, E. (2016). Introduction: Locating gentrification in the global east. *Urban Studies, 53*(3), 455–470. https://doi.org/10.1177/0042098015620337.

Smith, L. T. (1999). *Decolonizing methodologies: Research and indigenous peoples.* New York: Zed Books.

Smollett, E. W. (1985). Settlement systems in Bulgaria: Socialist planning for the integration of rural and urban life. In A. Southall, P. J. M. Nas, & G. Ansari (Eds.), *City and society: Studies in urban ethnicity, life-style and class.* Leiden: Institute of Cultural and Social Studies, University of Leiden.

Vangeli, A. (2017). China's engagement with the sixteen countries of central, east and Southeast Europe under the belt and road initiative. *China & World Economy, 25*(5), 101–124. https://doi.org/10.1111/cwe.12216.

Wiles, J. (2005). Conceptualizing place in the care of older people: The contributions of geographical gerontology. *Journal of Clinical Nursing, 14*, 100–108.

Zhou, Y. R. (2012). Space, time, and self: Rethinking aging in the contexts of immigration and transnationalism. *Journal of Aging Studies, 26*(3), 232–242. https://doi.org/10.1016/j.jaging.2012.02.002.

# INDEX

© The Author(s) 2020
D. Iossifova, *Translocal Ageing in the Global East*,
https://doi.org/10.1007/978-3-030-60823-1

143

Printed by Printforce, the Netherlands